RD LOAN

GEOGRAPHIES OF YOUNG PEOPLE

Anxieties over children's safety or teenage propensities towards violence and sex, combined with cases such as James Bulger in the UK and the Columbine and Santana school shootings in the US, precipitate moral panics in a large swathe of society. *Geographies of Young People* traces the changing scientific and societal notions of what it is to be a young person, and argues that there is a need to rethink how we view childhood spaces, child development, and the politics of growing up.

The book challenges popular myths that evoke general notions of childhood as a natural stage in the development towards adulthood. Alternative, contemporary psychoanalytic and feminist theories provide less structured perspectives, which value the embodiment and local embeddedness of young people. Enduring throughout is the conviction that a focus on children's geography and space is as important as history, because they articulate how places, institutions, and mechanistic notions of justice teach young people how to behave and why young people resist this kind of disciplining. The latter part of the book considers children's rights and suggests the need for a more nuanced and contextualized account of justice.

This book brings coherency to the growing field of children's geographies by arguing that although most of it does not prescribe solutions to the moral assault against young people, it none the less offers appropriate insights into difference and diversity, and how young people are constructed.

Stuart C. Aitken is Professor of Geography at San Diego State University.

Paperback cover photograph: This was taken by Stuart Aitken's friend, David MacLachlan, a Church of Scotland minister and social activist. During the Sandanista revolution, David toured northern Nicaragua distributing supplies and offering whatever help he could. He took the photograph of these two boys on his way out of a small town that had been attacked by Contras days previously, in search of a renegade leftist priest. His guide spoke of a possible new reprisal within the next few days. David was struck by the pathos of these two youngsters in the midst of a war zone, and his relative safety on a bus headed out of town.

CRITICAL GEOGRAPHIES

Edited by Tracey Skelton

Lecturer in International Studies, Nottingham Trent University

and Gill Valentine

Professor of Geography, the University of Sheffield

This series offers cutting-edge research organized into three themes of concepts, scale and transformations. It is aimed at upper-level undergraduates, research students and academics and will facilitate interdisciplinary engagement between geography and other social sciences. It provides a forum for the innovative and vibrant debates which span the broad spectrum of this discipline.

GEOGRAPHIES OF YOUNG PEOPLE

The morally contested spaces
of identity

Stuart C. Aitken

London and New York

First published 2001
by Routledge
11 New Fetter Lane, London EC4P 4EE

Simultaneously published in the USA and Canada
by Routledge
29 West 35th Street, New York, NY 10001

Routledge is an imprint of the Taylor & Francis Group

© 2001 Stuart C. Aitken

Typeset in Perpetua by BC Typesetting, Bristol
Printed and bound in Great Britain by
TJ International Ltd, Padstow

British Library Cataloguing in Publication Data
A catalogue record for this book is available from the British Library

Library of Congress Cataloging in Publication Data
Aitken, Stuart C.
Geographies of young people: the morally contested spaces of identity/
Stuart C. Aitken.
p.   cm. – (Critical geographies)
Includes bibliographical references and index.
1. Child development. 2. Environment and children.
3. Human geography. I. Title. II. Series.

HQ767.9.A38 2001
305.23–dc21   2001019311

ISBN 0–415–22394–6 (hbk)
ISBN 0–415–22395–4 (pbk)

In memory of Yumiko, who loved children
unconditionally

CONTENTS

FIGURES

PREFACE

Writing a preface is a curiously satisfying experience. It comes at the beginning of a book but it is almost always written when the rest of the text is done. Today, I tweaked the last words in my final paragraph and saved the file on a zip-drive to send to the publisher. That was very satisfying. But now I sit at home with this preface and I am taken – for this is what prefaces are supposed to do – to the beginnings of this book, to its inspirations, to the questions that began it all. Prefatory remarks are meant to initiate the reader to the ways writers came to think about the materials in a book without giving away too much about the central arguments. What is clear to me now is that the process of writing, in and of itself, helped precipitate a clearer understanding of my central arguments as they relate to how and where the current moral crises of childhood are constituted.

The idea behind this book was to revise a previously published resource monograph of the Association of American Geographers (Aitken 1994). When the resource monograph series was put to rest in 1998, I approached the AAG about obtaining the copyright for my work. They graciously returned the copyright to me and I convinced some editors at Routledge of the efficacy of a revised version of the manuscript with a new chapter or two highlighting changes in the study of children's geographies over the last decade. Therein lies the rather paltry inspiration for this work, but the final product bears very little resemblance to that inspiration. The book is no longer a recapitulation and extension of some earlier arguments with a review tacked on of some of the important work over the last decade. In the process of writing, I found myself using material in the original manuscript much less than I anticipated. And when the older material was used, I found myself using it in very different ways. The earlier monograph is of another time and place. My views today are not only more fully informed by feminist and post-structural theories, but they are sanctioned by a series of moral imperatives that I find particularly pertinent. It has taken the process of writing for me to discover these imperatives as they elaborate a series of questions about children's geographies that are disturbing. These questions focus on the nature of childhood, and children's experiences as they relate to their bodies, sexualities, ethnicities and their relations to adults. Ultimately, the questions push the social construction of childhood and children's experiences to a critical consideration of rights and systems of justice and therein lies the morality of the work. The questions are sometimes too threatening to put into words, but what has become clear to me through this writing is their centrality to understanding larger social and material transformations. The global implications of "putting children in their place" and a sensitivity to young people's local contexts play a large part in contriving current academic concerns for young people's well-being.

Some of these concerns were forefronted but not that well conceived when Doreen Mattingly, Tom Herman and I convened a National Science Foundation sponsored workshop in San Diego in November 1998. We invited a score of interdisciplinary scholars (some with established writing agendas and others in the midst of PhD research) with interests in children's geographies for a week of discussion about the evolving moral imperatives in the field. The published research and ideas of many of those workshop participants informs the current work. Our collective unease with standard categories of defining children and adolescents and normalized theories of development was founded on an engagement with new, less stable geographies and mappings, and a series of crises of representation that highlighted reflexivity and "being in the field" with children. The various crises of representation highlighted a moral turpitude that is embedded in changing conceptualizations of young people. Distinguishing children by ethnicity, sex, physical ableism or home context was also discussed because, at some point, continually shifting our attention to smaller arenas of difference suggests a hopeless relativism that borders on paralysis. So, what of studying children as a unitary category akin to the category "women" of the feminist movement? Certainly, there is an argument that children are similarly disempowered by the "patriarchal bargain." But it seems that distinguishing children simply as nonadults or young people is also unsatisfactory because it establishes a simplistic binary and directs attention away from the complexities of children's daily lives by emphasizing what children lack before becoming adults. Rather, we agreed that a fluidity of terms to describe kids and teens seems appropriate to their shifting identities. I construct young people in this way throughout the current text, but I do so not to denigrate important differences in the contexts of young people's lives. Rather, I try to point out the baggage (and disempowerment) that is associated with most categories and cohorts of understanding. Moral integrity revolves for me around knowing what questions to ask and what not to ask of these categories.

Although largely a thought piece, this book is also about working with young people. I make a point of omitting the voices of children from this text although the images of some of the young people I've worked with do appear. My altruistic concern is to empower young people; my selfish concern – less today than when I wrote the 1994 monograph – is to garner respect and propel my career forward while perhaps simultaneously presenting the impediments of some young people to social/spatial justice and achieving "the good life." But the reality of what Bill Bunge (1971, 170) calls the "immediacy" of fieldwork is that I forget about these disembodied issues – my career and my critical theory – when I'm spending time with children. The immediacy directs an embodiment in that moment with that child and it is all about our inter-personal relationship. I want them to like me, but it is also my hope that the immediacy of the fieldwork is liberatory in the sense that we get to play. Play encompasses the kinds of friendships I want with children, but it also embodies creativity and opens the practice of questioning. By so doing, it may encourage them at some other time and in some other place to project their voices and their positions. This is my hope. But, today, with this book, it is me who gets to play.

Stuart C. Aitken
San Diego
January 2001

ACKNOWLEDGMENTS

Many of the ideas for this book germinated in a workshop sponsored by the National Science Foundation (SES 97–32469 ''Young People's Geographies/ The Geographies of Young People'') organized by Doreen Mattingly, Tom Herman and myself. Thanks go to the participants of the workshop: Harold Bauder, Cindi Katz, Elsbeth Robson, Elizabeth Gagen, Gill Valentine, Sue Roberts, Herb Childress, Kimi Eisele, Melissa Hyams, Michael Baizerman, Sarah Holloway, Harriott Beazley, Sue Ruddick and Mary Thomas. I would also like to thank the many colleagues and students in Europe, Asia, Australia and North America who continue to encourage me with this work. But most of all, I want to thank my kids, Ross and Catherine, without whom none of this would be possible.

The author and publishers would like to thank the following for granting permission to reproduce photographs in this work:

Cindi Katz, City University of New York, for Figures 5.2 and 5.3.
London Free Press Collection of Photographic Negatives, D.B. Weldon Library, University of Western Ontario for Figure 1.1.
David Maclachlan for the cover photograph.

Every effort has been made to contact copyright holders for their permission to reprint material in this book. The publishers would be grateful to hear from any copyright holder who is not here acknowledged and will undertake to rectify any errors or omissions in future editions of this book.

1

PUTTING YOUNG PEOPLE IN THEIR PLACE

I begin this book with two stories that contextualize some of my interests in children's geographies. Although personal, I want to suggest that these stories reflect, as least in part, aspects of a recent sea-level change in ways social scientists encounter young people and theorize the fields of childhood and adolescent study. The first story takes place in a children's museum in Canada when I was a doctoral candidate; the second occurred in Mexico City about a year after I finished my PhD. They are stories about encounters with children but they are epiphanal because they changed, at an indistinct but fundamental level, how I think about young people and their place in the world. I present them here to first help establish some preliminary images for what I want to talk about in the balance of the book. They also enable me to think more clearly about how social science in general, and geography in particular, has in recent years come to know children in a critical and generous way. These new ways of knowing, as I see them, encompass an increased reflexivity between adult researchers and younger participants, they propel concerns about the discourses of science and nature within which childhood and adolescence fitfully repose, and they engage sensitive issues about the ways young people compose their identities. They also broaden scales of concern beyond children and their families with appropriate critiques of globalization, citizenship, young people's rights and the moral relations between young and old. The second part of this introductory chapter details this critique, and lays out a basis for some of the arguments I make in the rest of the book. Taken from these changes in how we come to know children and their place in the world, my arguments suggest a moral imperative to reconsider critically children's spaces as personally and politically embodied and locally embedded, and as a harbinger for new ways of understanding development.

Encountering young people 1: the cave

I sit on the plaster ledge of a simulated cave in a museum and scratch vigorously

Figure 1.1 Stuart-the-caveman

Source: Photo Archive for *The London Free Press*, J.J. Talman Regional Collection, D.B. Weldon Library, University of Western Ontario. Reprinted with permission

under my arm (the simulated animal skins are a bit itchy) before swinging onto the floor and busying myself with some bones strewn across the floor. I pile the bones up and then, unsatisfied, pull them apart before taking some pains to balance two or three of them in what I think is quite an interesting sculpture. My back is to the entrance of the cave enabling me to satisfactorily ignore my handler's entreaties to come and meet some new visitors. My handler, Lisa, is a spelunker, replete with hard hat and carbide lamp. Lisa is a real caver and a qualified museum interpreter who helps young children understand what it is like to live in caves. I am not a real caveman nor am I an interpreter, but today I am an important prop in Lisa's exhibit. She has brought a small group of young children (the oldest looks about five) to meet "Urg." That's me, it is an eponymous title because it is all I say. It is part of my performance as a 4,500-year-old cave dweller.

Earlier that year Lisa and I were with a group of cavers exploring Blowing River Cave in Tennessee. Preserved in the mud of this cavern are footprints of nine explorers (five men, two women and one child) who were probably looking for mineral salts. Here was an intrepid group of adults and a child exploring the reaches of a dark and inhospitable environment. Radio carbon dating of scattered charcoal believed to be from torches used by these early explorers dates their trip to around 2,500 BC. We used carbide lamps and two back-up sources of light. As I sat on a ledge in Blowing River with Lisa, I rustled my bag with extra carbide and wondered if those early explorers – if that is what they were – worried whether they had brought enough reeds for their torches. We discussed the immense feat of coming this far back into a cave with only reed torches and speculated on the possibility of another entrance. The winds of the cave suggested that if such an entrance had existed, it was long ago sealed by earth movements. We decided that a venture this far into the cave was near impossible without contemporary equipment. As I scurried along their route carefully avoiding the areas where footprints remained I wondered if the child felt the thrill of exploring places where few people dared go or if this for her was a common occurrence. I wondered at the arrogance of transferring my excitement, wonder and fear onto a child from 4,500 years ago whose footsteps I followed.

It is not my imagining of that neolithic child's story or my adventures in Blowing River that I want to relate here, but the ''Urg'' event that followed shortly thereafter. It is an event that none the less relates to the way I imagine relations with children. Some members of our trip decided to build a cave in the children's museum where Lisa worked. The cave boasted simulated limestone formations, rooms, crickets, bats and tight squeezes suitable for the passage of only the smallest bodies. Lisa developed an interpretative program and I volunteer to spend the weekend as Urg. The cave exhibit is ''hands on'' and I am to be played with.

The five youngsters cower behind Lisa as they enter the room of the cave where I am playing with the bones. Their parents and a newspaper photographer come in behind them. Lisa calls to me and I sniff the air. Then I slowly turn to face my audience. I can see that the children are interested in me but are too nervous to leave Lisa's protection. The photographer's camera flashes and I startle. Feigning terror in the face of this unknown disturbance, I scamper to the back of the cave and cringe behind a fake limestone formation. For some reason, my actions embolden two of the children who tentatively approach me and, taking my hand, lead me back into the center of the room. I am crouching at about the children's height as one throws her arms around my neck as if to ward off the advances of the photographer. The ensuing photo-opportunity is spoiled by another child who scolds the photographer for trying to frighten me. And suddenly I am theirs: Lisa and parents are forgotten as the children surround me with affectionate touches and protective body language.

I am not sure what I became in that simulated cave in my simulated animal skins – a pet, a plaything, a confidant, an ally against adults – but for the next half hour I felt like I was a trusted part of the world of those children. We played with the bones and then they took me out of the cave and showed me the rest of the museum. They explained how coke machines worked and that the frightening skeleton of an Albertosaurus wouldn't hurt me. They showed me some of the museum's exhibits and explained how they should be used. They got a kick out of using me to scare adults. I realized why the people in Goofy and Mickey suits at Disneyland have so much fun. I wondered if I, like the Disney characters, was simply another commodified pretext for what should fill a child's world. I was an adult construction of new ways that children should learn – hands-on, festive, fun, playful – about worlds that certainly did not exist the way we were portraying them. But there was something different here that went beyond education, museums or Disney. I was part of the play of children and their trust enfolded me in an enticing and carefree space of belonging. It was as if, by doing nothing of any great importance, we were doing the most important thing that the particular moment enabled. When the children reluctantly left I went back to my cave and waited eagerly for the next group.

My encounter with those young children is an indelible part of how I now approach the study of children. At the time of my museum performance I was beginning a dissertation that drew in part from cognitive behavioral geography. It was the early 1980s and prevailing methods for studying children were heavily influenced by Piaget's theorizing that children invariably experienced life through a structured series of developmental stages. Geographers used standard Piagetian bench tests, and developed sketch mapping and wayfinding exercises along with aerial photograph interpretation to help them establish the spatial efficacy of developmental stages. As interesting as these studies are, my encounters with children such as those in the cave suggested something more was needed to gain a broader understanding of children's geographies. The Piagetian paper and pencil tests seemed too far removed from children's lived experiences and they problematically positioned the researcher as an objective, impartial assessor of how well children performed. I found it quite disturbing that many of the papers on children's mapping and wayfinding abilities cited as inspiration Edward Tolman's (1948) work on "cognitive maps in rats and men." At the time I was less concerned about the bestial and sexual metaphors than the precise rationale and processes suggested by the model of learning and development espoused in these works. As I performed Urg in that simulated cave, I was engaging with children in a very serious enterprise and I knew that their worlds and their sense of wonder could not be understood through simulations and tests. The mapping metaphors were problematic because they challenged a seemingly sacred space and reduced it to the same kind of Euclidean dimensionality that found its most poignant applications in warfare and colonial expansion. It seemed to me that an agent of imperial conquest might well constitute an inappropriate tool for engaging young minds. My feelings about children's sketch maps rested fitfully against my training in cartography, but they came at a time of new maps, new geographies and new ways of thinking about space.

An understanding of maps wasn't the only thing to change in the 1980s. An epistemological and moral sea-level change swept many disciplines in the social sciences onto uncharted territories where traditional techniques and understandings precipitated inadequate and undesirable encounters with young people. Theories and methods assumed to be unassailable during my years in graduate school no longer retained any kind of solid semblance. Some of us became unsure of how to approach our subjects, and even less clear about how to situate ourselves in what we studied. We did not want to be distant, impartial laboratory technicians or objective field researchers who used metaphors of rat behavior and cognitive maps to understand children and their worlds.

A much heralded "crisis of representation" un-moored academic practices of writing and this crisis became particularly appropriate for understanding where we situate ourselves in relation to young people (Yuval-Davis 1993; Stephens 1995). Terms such as "children" and "youth," heretofore thought unproblematic, became sources of significant debate (Griffin 1993). Are they naturalized as the lower (less than) part of a developmental sequence? Why is the term "kid" acceptable to some youngsters while "child" suggests different connotations? At what age does childhood begin and end? How do we come to know adolescence? In contemporary usage there is constant slippage between the terms that characterize young people. Sometimes, as in the case of the United Nations Children's Fund (UNICEF) and The International Program on the Elimination of Child Labor (IPEC), the term child is used to describe all people under eighteen years of age. In other contexts child is used to describe pre-pubescent young people, with terms such as adolescent or teenager used to describe older children. At one level, the state defines entry into adulthood but, at other levels of culture and commerce, there is a homogenization of symbols to the extent that it is difficult to know where the child ends and the adult begins. For some, being a child is opposite to being an adult, and this conceptualization is sometimes extended to suggest that childhood is simply the absence of adulthood. For others there is recognition of the particular qualities of childhood, but this is tempered by the belief that with maturity come distinct improvements. This is certainly the case with spatial cognition and developmental theories, which suggest that through certain stages children gain the maturity to handle more complex environments and ways of knowing.

The concepts of childhood and adolescence are necessarily linked to that of adulthood but *adolescence* complicates these webs of meaning. The widespread use of this term is relatively new, and the fact that it is regarded as an appropriate category for scientific study may be attributed to the influential work early this century of Granville Stanley Hall (1904). But today most academics recognize that distinctions between children, teenagers and adults are politically contrived. Gill Valentine, Tracey Skelton and Deborah Chambers (1998, 4) point out that much of the scientific concern derives from "anxiety about the undisciplined and unruly nature of young people (particularly working class youth) [that] has been repeatedly mobilized in definitions of youth and youth cultures for over 150 years." Indeed, Valentine (1996, 582) notes that in interviews with parents about issues of safety, many frequently "elided their discussions of the under 12s and teenagers when talking about 'dangerous children.'" Sharon Stephens (1995, 13) argues that although there is a growing consciousness of children *at risk*, there is also a growing sense of children as *the risk*. In stories of street-children from Rio de Janeiro to Los Angeles, children are represented as malicious predators who are unshackled from moral and social responsibilities. Recent research suggests larger institutional and societal constraints, and a quixotic search for identity by those who are young and homeless. Harriott Beazley (2000, 2001), for example, identifies "geographies of resistance" in Yogyakarta, Indonesia, where a response to the larger "spatial apartheid" of the State empowers street-children. In her study of seemingly dangerous youth in Los Angeles, Sue Ruddick (1995, 1998) argues that terms such as delinquent, punk and runaway are conflated in apodicitic geographies that disenfranchise homeless young people from the spaces that serve their identity needs. That this crisis takes place simutaneously in Yogyakarta and in Los Angeles is an important point. As Stephens (1995, 8) points out, the "[c]urrent crises in notions of childhood, the experiences of children, and the sociology of childhood are related to profound changes in a now globalized modernity in which the child was previously located." Children experience the world as it is manifest, and they are part of the ideology, war and corporate greed that is exported from the boardrooms and strategic command centers of Washington, Paris and London. It is within this context that I place the photograph on this book's paperback cover. Taken by a friend of mine who worked in northern Nicaragua during the Sandanista revolution, it depicts two boys "doing nothing of any great importance" in a highly politicized global space. It is a space that was contorted by greed, ideology and racism. Contras looking for a leftist priest days previously had attacked the boys' village. The boys hang out for a moment while the village cringes before the possibility of another attack. Mean streets, war zones, public kid corrals and private family realms of child abuse. These spaces are the places of children, and they embody the pretensions of our adult sexed, raced, politicized and able-bodied selves.

The plasticity of terms such as child and adolescent begs for more nuances than offered by traditional educational and developmental theories. The reason for this need, I would argue, is because of the moral turpitude that is embedded in changing conceptualizations of young people. Distinguishing children as non-adults or young people is also unsatisfactory because it establishes a simplistic binary and directs attention away from children's daily lives by emphasizing what children lack before becoming adults. The fluidity of terms to describe kids and teens seems appropriate to their shifting identities and so I make no excuses for, indeed I make a point of, slipping between concepts such as infant, toddler, youth, child, adolescent and teenager. I do so not to denigrate the important differences between toddlers and adolescents but to point out the baggage (and disempowerment) that is associated with these terms. What particular kind of bodily comportment does the term toddler suggest? What does it mean, for example, to call a 13-year-old a teenager, a gangly youth, an adolescent or a pubescent child?

Nor am I suggesting with a focus on those who are "not adults" that a universal category of experience exists. Although children are defined in relation to adults, as Holloway and Valentine (2000b, 6) point out "other differences also fracture (and are fractured by) these adult–child relations." Children grow into the pretension of identities that reflect race, class, gender, bodily appearance and other socially constructed differences. These axes of difference are not singular or additive but they constitute rather multiple transgressive and transformative features of identity. My calling upon a series of "adult, Western views" in this book may suggest an essential and universal notion of "the child," or that children are passive in creating their identities and that their lives can only gain meaning through adult values. This is clearly not the case, but perhaps it is worthwhile thinking of the terms used to describe young people as parasitic because they often inculcate children as research subjects, or as part of the family realm, or they commodify their desires into a form that garners maximum profit for global corporations. This book is about the social construction of young people through time and in space. It is about the Western ideological construction of childhood as a privileged private domain of innocence, spontaneity, play, freedom and emotion in opposition to a public culture of culpability, discipline, work, constraint and rationality. There is a growing body of work on Western childhood suggesting the creation and expansion of the modern antimony of child/adult, like the female/male dichotomy, is crucial to setting up hierarchical relations upon which modern capitalism and the modern nation-state depends (Postman 1982; Qvortrup 1985; Stephens 1995). In what sense does this antimony help engender a status quo that reduces the lives and practices of young people to something from which corporations can profit? Put another way, how are the constraints on young people's lives also about how to make adult lives more comfortable?

Concerns about how the lives of young people are constrained by contemporary Western values are sometimes paralleled in social science research by how they are *othered*. Characterizing young people as other is a particularly thorny problem. Of all people who are constituted as other in that they are different from us, young people are particularly perplexing because they are intimately part of our lives and they are, at least in part, constituted by what we are and what we do. Perhaps this is why children are not always "othered" with the same exclusionary ferocity that permeates some discourses on women or people of color but, rather, they are openly valued (Holloway and Valentine 2000b, 4). To add to the problem, adults all used to be young and are sometimes predisposed to make claims about knowing what the lives of young people are about. What biographical baggage do we, as researchers and adults, bring to the study of children and youths? To what extent can our adult position invalidate our ability to empathize and situate ourselves authentically in the lives of children? Children see things in environments that we may have forgotten how to see, let alone understand. If we can no longer fully empathize with young people or imagine what they experience, how can we write for them and establish agendas on their behalf? How do we perform our lives and our research around young people? What do we learn about children's geographies from the performances of Jean Piaget, B.F. Skinner, Pee-wee Herman or Stuart-the-caveman? And what about the performances of children?

Encountering young people 2: the Zócalo

The music they perform is loud, discordant and, if you choose to see and hear them in that way, they are precocious and disrespectful. I can hear them from the other side of Mexico City's central plaza, the Zócalo. They are very loud; they want to be heard. I approach and watch them, listening to their busking through several chaotic, unrecognizable tunes. The oldest boy plays trumpet while a younger boy and girl thrash a snare and a base drum respectively. The children seemed intent upon their music rather than the collection bowl in front of a baby who sucks her fingers seemingly oblivious to the cacophony. This is a hallowed place, but the four seem too absorbed in their music to notice the *genius loci* of the Zócalo. In the background, part of the great Aztec pyramid of Tenochtitlán is in the process of being excavated and recreated. In the foreground, the buskers play to the ornate side of the huge cathedral which dominates the plaza.

The cathedral, built by the Spanish from the ruined stones of the original Tenochtitlán pyramid, represents the high culture of a conquering people. The recreation of Tenochtitlán's pyramid is an attempt to rebuild the material culture of an indigenous civilization that was torn apart and reduced to rubble. It is a simulacra recreated to represent nostalgic notions of an earlier historical moment. Sandwiched between these two sad monoliths of cultural politics, the children perform. They want to be heard.

The bustle and noise of the city ebbs and flows around the cacophony. Someone digs into a pocket for a peso or two in sympathy rather than from appreciation of

Figure 1.2 An image composition incorporating Mexico City's Spanish cathedral
(left), the simulacra of the Aztec pyramid of Tenochtitlán (right) and the
children of the Zócalo (ghost-like, top center)

the children's rendering of musical styles. I recognize little beyond the passion of
their offense although, on occasion, I hear the beginnings of a jazz theme or the
kernels of a rock 'n' roll beat. Just as quickly as a musical structure appears, it
dissolves into something new and incomprehensible. Although the children's
chaos of notes and rhythms is impossible to comprehend, I stand listening and
watching for quite some time. It seems to me that the loud and undisciplined
music de-stabilizes the heart of this, one of the largest urban concentrations of
humanity on earth. The old Spanish cathedral and the recreated remains of
Tenochtitlán's pyramid seem unable to deal with the intensity and passion
of the youngsters' performance. And yet these two structures remained stoic
monoliths to the contradiction of cultural expression and cultural oppression.
For me, they represent part of an insidious historical and geographical global
structure that contextualizes, constrains and exploits the lives of young people.
Here, in stark relief, are icons to an old, dark colonial order (built from a
ruined but equally tyrannical empire) alongside the recreation of an authentic
and prosperous Aztec past poised to realize tourist dollars and new rounds of

capital investment. Squeezed between these icons of imperialism, with discordant melodies and dissonant rhythms, the children of the Zócalo contest an adult space of past and present cultures.

The children's rage against imperialism is my romantic image, and it diminishes their dire poverty and their economic will to survive. I am conscious of my distorted rendering. The pesos collected in the bowl in front of the baby map these youngsters onto the productive activities of Mexico City's informal economy. The bowl speaks volumes about a lowly scale of production in a global economy rife with neo-liberal free trade ideals; it confuses IPEC discussions about what constitutes child labor.

It appears that many would-be patrons avoid the musical assault of the children, preferring an alternative route across the plaza. It occurs to me that the children are oblivious to the cathedral, Tenochtitlán, and the passers-by, they are so engrossed in their toil. The children are clearly enjoying their labor and are heedless of the extremely slow pace at which their bowl is filling. At that moment, it seems to me that their actions are not really about economic survival. Nor do they care about what we hear. These children are not playing music for our edification or for monetary remuneration, they are simply *playing*. But by doing so in such a place as this they call into question, for me at least, the ways whereby children's everyday lives are structured, constrained and exploited by larger global forces.

They highlight a further question that recalls my earlier question about othering and the context of the two boys on the front cover, sitting under the "Viva Sandino" proclamation. How do I locate those boys and the children of the Zócalo "in their own terms"? Although you may have heard the music of these children, you do not hear their voices. Nowhere in this book do the voices of children appear. I am struck of late by the attempts to include the voices and views of children in some of my colleagues' work. Some of these attempts are laudable but more often than not – particularly when younger children are involved – they are disingenuous and, at best, a form of tokenism. But what of the differing contexts of children in Mexico City, Managua City and Yogyakarta? Beyond discussion of some key works in the global south, my focus here is on the contexts of childhood as it emanates from the global north. Indeed, the examples I pull from the global south are used to suggest that the hegemonized position of young people is derived from the global north. But, as Stephens points out, the cultural contexts of childhood should not be bought at the cost of an awareness of the complexities of cultural definition elsewhere:

> Rather than merely explicating Western constructions of childhood, to be filled out in terms of gender, race, and class differences and to be compared with childhoods of other cultures, we need to explore the

global processes that are currently transforming gender, race, class, culture – and, by no means least of all, childhood itself.

(Stephens 1995, 7)

Understanding the global complexities and differences in children's experiences is important but beyond the scope of this book. My goal is a more modest explication of global processes as they are linked to the nature of childhood, bodies, sexualities and the changing spaces of children. In a world of shifting boundaries there is perhaps at some level an obsession with the boundedness of the body, the family, the community and the nation and perhaps also an increasing inability to cope with children whose unchildlike behaviors and attitudes suggest that they will not fulfill expected traditional roles. My hope is to elaborate the different sides of the spatiality that constitutes the construction of young people's identities – their embodiedness and embeddedness – and to relate this to the moral geographies that emanate from these kinds of adult concerns. On the one hand, I am concerned with the endless carving up and claiming of space by different individuals and constituencies and how that process disenfranchises young people. On the other hand, I am concerned about the connectedness of all spaces, and all cultures, even those at the other side of the world.

Concepts such as "the global child" and "the world's children" emerged in the official discourses of international agencies such as UNICEF, ILO (International Labor Organization) and WHO (World Health Organization) in the late 1970s. Depictions of children's lives in the global south and in central cities in the global north suggested a problematic distinction between Western concepts of what childhood should be and the actuality of some children's lived experiences. The issue here is not that the diverse lives of children was never noticed prior to this time, but that the deviance from the Western norm could be interpreted not as local particularities but as instances of backwardness and underdevelopment, thus justifying expanded efforts to export modern childhood around the world (Stephens 1995, 19). Since that time, terror tales emanating from the global media of child abuse – particularly child sexual abuse – further challenges traditional beliefs of what constitutes childhood. We learned that child abuse was not just a problem of the global south but resided also in neighborhoods close by. Curiously, efforts to quell the problem focused primarily on education about public-focused "stranger danger" and educational programs targeting low income neighborhoods. Again, a myth of modern childhood garnered from the private realm of white middle-class ideals is propelled into public global spaces.

Stephens (1995, 8) was the first to argue that the global crises of childhood are not so much sociological and psychological, but rather temporal and spatial. An historical perspective on "the world's children," she argues, suggests complex globalizations of once localized Western constructions of childhood and adolescence. To identify current crises in the sociology of childhood and the experiences of young people requires a focus on processes that are related profoundly to changes in a globalized modernity in which "the child" was previously located. It is to an understanding of these Western processes that this book is addressed, and it is from fundamentally geographic perspectives that it gains insight.

Geography and the study of children

> Children are not selected because they are a safe subject. They are a furious subject: the most furious subject.
>
> (Bunge and Bordessa 1975, 1)

An articulation of moral geographies for young people can be traced back to the early 1970s. Born out of increasing quantification in geography and a search for scientific theory in the 1970s, Bill Bunge's (1973; Bunge and Bordessa 1975) geographical "expeditions" in Detroit and Toronto focused upon the spatial oppression of children. The central thesis of this work posits children as the ultimate victims of the political, social and economic forces that contrive the geography of the built environment. Starting with observations of children at play in inner city neighborhoods, Bunge's expeditions employed a myriad quantitative and qualitative, aggregate and individualistic geographic approaches to the study of spatial structure and interaction without losing sight of the central theme of children's oppression.

Cutting his academic teeth as an apocalyptic disciple of the quantitative revolution in geography, Bunge was one of the first to express his discontent with other messengers who, he felt, capitulated to capitalism (Merrifield 1995, 49). With a specific focus on spatial oppression, he scorned quantitative geography's concern with economic modeling and regional science. Despite his disaffection with the product of geographic science, Bunge remained committed to spatial interaction, traditional mapping and the moral certitude of logical positivism: "I am a scientist. That is enough. I do not make value judgements in my work or worry about ethics and doing right (or wrong). I simply do science, geography" (Bunge 1979, 172). Although seen as problematic from the standpoint of contemporary critical theory, Bunge used the seeming virtue of science to draw attention to what he thought was a moral crisis in society. That crisis was reflected, he argued, in the plight of children and it evoked Darwinism as a corollary to the importance of keeping children safe. The pressure of the environment on the young is crucial to any species surviving, Bunge averred, because it suggests that the "Darwinian logic of child protection, not sentimentality, not kindness, not value judgment, turns the search for survival, into a children's book'' (Bunge and Bordessa 1975, 2). Like canaries in a coalmine, for Bunge, children were a barometer to measure the wellness of society and spatial statistics revealed the patterns of that sickness. Information garnered from the expeditions gave sustenance to political programs when presented to city politicians and planners. Bunge's report on school decentralization in Detroit, for example, highlighted hunger and the expense of bus fares to show the very real difficulties low-income African-Americans faced in simply showing up for classes. Geographers and local community leaders participated in the report to produce a series of maps indicating a more appropriate and socially just geographical allocation of educational resources (Merrifield 1995, 56). Elsewhere, empirical data were teased and cajoled into propaganda maps of Detroit and Toronto that displayed infant mortality, doctorless regions, toyless landscapes, grassless spaces, machine landscapes and rat-bitten baby landscapes.

In one example, Bunge used the relationship between high-rises and children to suggest that with their dearth of play space these vertical edifices of machine architecture forced children onto the dangerous machine spaces of city streets. The expedition's rhetoric – children caged and buried in the sky (Bunge and Bordessa 1975, 76) – is as evocative as the example is emotive. At a time in geography when large aggregate studies sought statistical norms, Bunge chose two 9-year-old girls residing in the same neighborhood who lived in low-rise and high-rise environments respectively. He hypothesized simply that the low-rise child would have an advantage over play and learning in the local area. Untestable from a statistical standpoint, a comparison of sketch mapping exercises revealed a home range that was extensive and planner for the low-rise child and intensive and pictorial for the high-rise child. Interviews suggested that the high-rise girl played primarily within the building and the confines of the parking lot. This contrasted with the ''horizontal flow'' of the low-rise child who played up to six blocks from home. The ''vertical flow'' of the high-rise child resulted in an extreme distance separation from parental supervision before the child could access the outside environment. The actions of a child on a street below are not easily followed by an anxious parent.

As part of his geographical expeditions, Bunge merged mappable data with extensive fieldwork in what Andy Merrifield (1995, 50) argues is the epitome of a search for situated knowledge because it offers ''a conceptual platform from which to call into question all privileged knowledge claims.'' Knowledge of this kind enables the making of new kinds of maps – Donna Haraway (1991, 191) calls these ''maps of consciousness'' – for people who by virtue of their class, ethnicity, sex or age are maginalized through masculinist, racist and colonialist domination. Situated knowledge is embedded in time and space and embodied in people and their actions.

13

For Bunge, the most important kinds of situated knowledge reside as part of the sphere of reproduction. In 1977, he published a review of the Detroit/Toronto geographical expeditions under the title ''The point of reproduction: a second front.'' The paper is a manifesto in which he argues for a focus on class, gender and race as issues of reproduction that are constituted in homes, families and neighborhoods. Bunge's first concern is that most economic geography is prejudiced against the ''non-working working class'' because it focuses on workplaces and the point of production. The ''hidden landscape'' of the home, he claims, is a legitimate arena for geographic inquiry as are the worlds of children's welfare, care and play. The point of reproduction is a point where marginalized groups – children, injured workers, retired workers, unemployed workers and sexually, racially, ethnically and religiously discriminated-against workers – constitute the majority of the working class even during periods of ''full employment'' (Bunge 1977, 61). Bunge's agenda was not only to find these groups but to ''establish their geography, their perceptions of space,'' so as to make ''their rightful claim to their turf'' (*Expedition Field Manual,* quoted in Merrifield 1995, 58). In a sentiment that presages current concerns for globalization and local practices, he urged that global problems be brought down to the scale of people's normal everyday lives.

At about the same time that Bunge was initiating his geographical expeditions, a fellow Marxist, Jim Blaut, co-founded with David Stea the ''Place Perception Project'' at Clark University (Blaut *et al.* 1970; Blaut and Stea 1971, 1974). With less polemical work than Bunge's, Blaut established an agenda for research in children's geographical learning, generated new theory and conducted some provocative empirical research on the early mapping abilities of children. Focusing on children's abilities and perceptions, Blaut's work was to lead to a focus in geography upon child development and spatial cognition (Moore 1976; Matthews 1984a, 1984b, 1986; Golledge *et al.* 1985; Downs *et al.* 1988; Blaut 1991; Downs and Liben 1991). Some of this interest stemmed from long-standing needs to understand how space and place emerge as important parts of the way the world is apprehended, and following from Blaut's work academic concern embraced how this understanding evolves from birth. Spatial cognition and developmental theory spurred interests in this area but often circumscribed the questions asked by geographers into a framework that constructs a universal child and quite specific, instrumental ways of knowing. In this it differed markedly from Bunge's focus on situated knowledge.

Bunge's expeditions also influenced rich empirical research in geography that focused on a humanistic, quasi-ethnographic documentation of children's and teenagers' interactions with their everyday spaces. In the years that followed the ''Place Perception Project,'' students from Clark University provided insights that enriched scholarly understanding of children's experiences of social reproduction (Ward 1978, 1988; Hart 1979; Wood 1982, 1985a, 1985b; Moore 1986). The most important distinction between these ethnographic methods and those of cognitive and spatial science is the relationship between the researcher and the researched. The ethnographer is not seen as an expert on the subject under study but instead the participant's expert knowledge of her or his culture is stressed. The first study of this kind in geography was Roger Hart's (1979) descriptive developmental study of children's place experience in a small New England town. By living in the town for an extended period of time and using a multiplicity of methods including direct observation, structured discussions and ethnographic interviews he was able to take an integrative look at the outdoor geographies of all the town's children. In particular, he noted the ways children appropriated and named public spaces – "the snow slide to the school bus," "the house with the dog that bites" – to the extent that they were imbued with child-like intentions and purposes. As his research progressed, Hart (1979, 105) began to reflect on social class differences in the way children played and in related differences in ideologies and philosophies of child-rearing. For example, he questioned the use of different yard toys as indicative of adult status and class. In an unsystematic way that can only be uncovered through a *bricolage* of ethnographic methods, Hart began observing the use of the environment as a tool in socialization.

Katz (1986, 1991b) takes up this exploration of the role of the physical environment in social and cultural reproduction in her study of children in rural Sudan. Although she uses many of the methods pioneered by Hart, Katz's work is a significant extension because she is sensitive not only to the role of the physical environment in social and cultural reproduction but also to the connection between space, place and power in terms of how children's local lived experiences are affected by global economic restructuring and change. In the case of rural Sudan, this involved reflection on the changes in the everyday lives of girls and boys after an irrigation scheme had altered the type of agriculture practiced in the village she was studying. In a further study Katz (1993) compares the local lived experiences of children in Sudan and North America in terms of the same restructuring processes. I will return to Katz's work on children and globalization in Chapter 5, but it is important to note at this point the value of ethnomethodoligical approaches that articulate a greater appreciation for the nuanced links between the physical environment and social and cultural reproduction. These methods rapidly eclipsed traditional developmental techniques that separated out the effects of the environment and tended to reduce children to psychological phenomena.

Denis Wood was one of the first geographers to articulate a sense of how the built environment and physical spaces inculcate larger global processes that impinge in hidden ways on the lives of children. In the 1980s, he wrote a series of provocative studies on children "doing nothing" that began a life-long interest in the hidden spaces of childhood (Wood 1982, 1985a, 1985b). His initial fieldwork in Barranquitos, Puerto Rico, comprised places very similar to what Kevin Lynch (1977) described as boring for children. Wood's methodologies, however, were far different from the cognitive mapping approaches of Lynch. Wood and his wife lived in the town for a year conducting in-depth ethnographically styled research. Focusing on children "doing nothing," he points out that children can also be bored "doing something" such as playing baseball. Rather than boredom, he avers, doing nothing is a searching, a time of change, a time of aesthetic:

> It has nothing to do with being alone. A lot of kids can be empty together . . . all activity [in Barranquitos] sprang from doing nothing. Sitting on the step before the stairs or standing around on the porch of Angel's *comado* (general store), the kids with nothing to do were poets waiting for their muse. . . . Doing nothing is filling. Doing nothing is an unfolding of things to do, an unfolding of things that have no names, like mooning around a lamppost or kicking stones into the drain across the street; an unfolding of things to do that do have names, like a whole string of street games such as *El Gato y Rarton* and *Escondido* and *El Pote*; an unfolding of things that have names that can as well be left unsaid like stealing and seeing who can pee the farthest; or an unfolding of all these things mixed together. Doing nothing is almost everything. As a term, it conceals as it identifies. It is both comprehensive and evasive, simultaneously screen and mirror. Like a kaleidoscope, it is everything, and it is nothing. Most of all it is *doing*. And it is what the Barranquitos kids did most of the time.
>
> (Wood 1985a, 9)

Of course, this kind of doing requires a certain degree of freedom. Children must have time to do nothing and the space within which to do it. They must have the freedom to take off down the street when they see something interesting going on. They must be able to hang out on the corner in relative safety. And they must be able to do all these things in private, on their own, removed from adult supervision. When a child is doing something (swimming in the public pool) it is usually sanctioned by adult authority and structured to be unresponsive to individual whims and fancies (no running or jumping in the pool), but when a child is doing nothing (skinny-dipping in the local quarry) there is no adult control and more child perniciousness, precociousness and personality. It may be argued that the freedom to be unsupervised and do nothing is becoming less and less a possibility for children, particularly in the global north. Non-activities are

16

seemingly inconsequential exchanges but they portend an important aspect of spaces that should not be challenged and controlled by an adult world. Wood's work points to the significance of maintaining portions of children's lives that are not organized and institutionalized by adults, which are not circumscribed by video games and store-bought toys, or arranged through "little league" tournaments.

Evidence from Wood's early work is impressionistic, but it suggests that as children grow up to and through adolescence their lives are circumscribed by a tension between activities that are organized and institutionalized by adults and the desire to find their own way in their own places. In later work, Wood's fieldwork focused on the controlled world of his children as contextualized by his home in Chapel Hill, North Carolina. He argues that the child's built environment – crib, room, home – is a social construction that defines lived experience by providing a forum through which sensations can be contextualized. The bounded space of the home provides security, comfort and familiarity, as well as protection against danger and the indeterminacy of public places. In collaboration with a linguistic psychologist, Robert Beck, Wood writes about his home comprising a field of rules composed of values and meanings:

> The pretense to universality crumbles under the demands of every specific site – culture is concrete: It is manifested not *in general* but necessarily *in situ* – and, although people *say* rules, they are *embodied* in specific actions and things. To enter a room is to find oneself immediately amid objects whose character and arrangement admit only of certain possibilities; it is always to enter a unique system of rules. The rules educed from this room may often be exogenous (most will be), but inevitably some will prove to be *sui generis*, and the ensemble will in any event be a singular property of the time and the place.
>
> (Wood and Beck 1990, 3)

Rules are embodied and encoded within a home's physical elements so that seemingly neutral space can be understood as a stimulus that is also a transformation of one or more "voices" (such as capitalism, social constructions of gender, or concepts of social class). These voices operate at the intended level of action, the child.

In a later book, *The Home Rules,* Wood and Beck (1994, xvi) elaborate more fully on the child as the level of action. The room or house is a field of rules comprising a myriad values and meanings embodied in ceilings, floors, doors and house plants. Adults who live in the house respond intuitively to these rules (as, more often than not, do visitors), but the presence of kids makes

them explicit. What for children is a field of rules is for the adult a nest of comforts. For adults, rules are hidden in images of Western comfort to the extent that
most appear uncontrived. But not only is comfort the consequent of a massive
contrivance, its means and ends are neither simple, straightforward nor direct
(Wood and Beck 1994, 26). The image of domestic propriety is that of a
mother, a father and their children lounging comfortably in a heated room,
perhaps with a fire, as the snow falls outside. The windows and doors must be
kept closed to maintain the heat, but this image of comfort is contrived at great
expense. The hidden voices that comfort requires – from the electricity provided
by the fission of uranium to the presence of American warships in the Persian Gulf
to protect supplies of oil – is a system of contrivances that extends "into every
pit and plain of the planet's economy" (Wood and Beck 1990, 13). The structures
of meaning (rules for kids and comforts for adults) do not originate in the room
but participate in larger structures of the world. This elaboration of the rules
embodied in the built environment complicates the social world by insisting
"on the inner contradictions that render it slightly bitter even at its sweetest,
to insist that it is not something one can have, that it is only something one
can be" (Wood and Beck 1990, 14). It is this larger designation of young people's
worlds and their inner contradictions that drives much of the currency in studying
the geographies of young people.

The energy propelling contemporary geographic interest comes from different
sources from those that inspired Bunge, Wood and Hart. Contemporary research
is perhaps most influenced by critical feminist and post-structural theories. For the
most part, these ways of knowing form a critical and reflexive engagement with
the lives of young people. Reflexivity and critical theory focus on positionalities,
playfulness and prescriptions for democracy, justice and the celebration of
difference. In the balance of this chapter I set out some of the arguments that
propel my concerns for the current moral assault on young people and the
ways critical geography provides appropriate contexts for unpacking this assault.
What follows is not meant as a rehearsal for the discussions that comprise the rest
of the book, nor is it meant as a summary of the topics covered in the book.
Rather, I try to articulate what I see as the central issues that help establish critical
children's geographies. These issues inform rather than stage the discussions that
follow.

Critical children's geographies

In 1994, I ended a monograph entitled *Putting Children in Their Place* with the
admonition that we must seek the child within ourselves. I did so not out of
any kind of romantic, New Age or Freudian enthusiasm for innocence or
repressed knowledge, but because of the possibility that in adulthood we separate

ourselves off from places and people to the extent that we no longer know how to *play*. I used play in a very specific way to mean engaging in dialogue with others, our environment and ourselves in much the way that Wood prescribes when he talks about elaborating upon inner contradictions. The counsel of Donald W. Winnicott (1971, 1975, 1988), a British psychoanalyst who had quite a lot to say about play, is that at some point in the making of our adult selves and our adult societies, we lose the ability to dialogue, embrace contradictions and maintain ongoing conversations. I was quite taken by this thought but it was one of those moments of striking insight that later I had to abandon, at least partially. What I specifically abandon in this book is the notion that in the passage from childhood to adulthood there is some irretrievable loss. I think things are significantly more complex than that. Indeed, my thinking on this resonates with that of David Archard (1993, 23), who suggests that ''it may well be our judgements as to what matters in being an adult which explain why we have the particular conception of childhood we do.'' The first thing I want to do with this book is tease out how the study of children is influenced by past thinking by adults to the extent that it is difficult to move beyond conventional wisdom on children embodied by science, developmental theory and concepts of nature. And so the next chapter focuses on the rise of the scientific study of children and its relationship to the study of nature with an evolving focus on development and pedagogy. Chapters 3 and 4 continue this discussion with a specific focus on the study of children's bodies and their sexuality.

Geographers, like other scholars, generally use children for different purposes and so it is necessary to discuss first how geography approaches these matters before considering some geographic material in the last three chapters of the book that perhaps shed more direct light on the matter. And so, in Chapter 5, I argue that a fuller understanding of young people's geographies is not only an important disciplinary pursuit but constitutes an important hinge-pin around which larger social transformations revolve. In Chapter 6 the notion of the material conditions of children's lived experiences is elaborated upon with consideration of the ways unchildlike behaviors challenge the norms of childhood and adolescence discussed in previous chapters. In the final chapter, I elaborate on contemporary notions of children's rights and suggest that new forms of justice are required to encompass an embedded and embodied notion of young people. My final arguments draw heavily from a neo-Winnicottian view of play and justice that relates to children's public and private rights.

In the balance of this chapter I elaborate upon the main themes of the book – nature, bodies, sexualities, social transformations, material conditions and justice – as they relate to the ways young people are *placed*, at what *scale* they operate and in which ways their identities are *fixed*.

Placed

I write in an era of profound change in how we come to know the spaces and places of young people. Geographers pose questions about the ways adult presence and absence is inculcated in those spaces and places. Childhood and adolescence are scrutinized and politicized by new theories that shed light on a series of interlocking spaces and interdependent places. Numerous empirically based studies challenge essentialist notions of a monolithic "child" by arguing that childhood is not only constructed in different ways at different times but also varies depending upon where it is constructed (cf. Holloway and Valentine 2000a). Analysis is complicated further by interpretative lenses that focus simultaneously on local and global representations and lived experiences. Traditional perspectives of childhood and adolescence as developmental and social classes are radically undermined, and we are left with perspectives that valorize difference and a clearer notion of the constraints imposed by socially constructed sets of meaning.

Recent geographic work is about the practices of young people, their communities, and the places and institutions that shape (and are shaped by) their lives. It is about quirky local geographies that spiral children into despair and violence, it is about youth movements and communities of rage and mayhem, and it is about omnipresent global geographies with uneven outcomes that conspire to commodify and exploit the lives of young people. Places are important for young people because these contexts play a large part in constructing and constraining dreams and practices. Although important points of convergence, local contexts may complicate our understanding of young people's geographies and highlight relativism to the extent that there is nothing critical to say. Through our study of local places we should not be silent about larger concerns. This book is about how we place children, but it does not dwell on the abiding (and important) complexities of local comparisons. Rather, I seek to elaborate larger concerns about the changing nature of childhood, adult moralities, young people's experiences and the embodiment of their political identities as they are embedded in the processes of globalization. How are children and youths contextualized in representations of global economic space and in what ways is it possible for them to attain cultural and social capital that has some power? Some of the recent work on children's geographies is influenced by what Doreen Massey (1993, 1994) calls a "progressive sense of place" because it is not circumscribed by the local, but rather extends discussion to the power that ebbs and flows from complex global connections. The notion of what constitutes a "normal childhood" (e.g. innocent, playful, carefree and focused on education rather than paid labor) from a wide range of studies in the global north are now challenged by work elsewhere that contextualizes children's work and play in different ways. These studies of children's varied places focus specifically on their day-to-day lived experiences in local contexts, but there is more to this research than simple comparisons.

Something more is indicated by speaking of childhood and adolescence rather than simply children and young people. The former are placeless and abstract nouns which denote the state of being a child or young person. They suggest a certain formal and sophisticated understanding of what and when it is to be a child or teenager, one that abstracts from the particularities of day-to-day lived experiences. If a society is to possess a *concept* of childhood and adolescence rather than an *awareness* of young people, it is likely to be informed at some level by theory (Archard 1993, 17). In the next chapter, I articulate the progressive articulation of theories of human development that inform some predominant contemporary understandings of childhood and adolescence. They are Western themes tied to Western literature, but they hold consequential sway on the context of "the world's children." I argue that the importance of Western scientific and literary discources on children and childhood from the beginning of the industrial era suggest more complex and critical positions in the construction of modernity than a model of young people simply as raw materials for the capitalist marketplace. Within an evolving discourse on the modern nature of childhood is hidden speculation on the body, sexuality and the construction of a racialized self, ideas that are central to the construction and export of a globalized capitalist ethic. The discussion in Chapter 2 on scientific and biological discourse, then, is a precursor to Chapters 3 and 4 where I argue that the spatial separation of young people from the adult world is not so much about material segregation but rather it is about enforced exclusions that comprise disembodiedness. These exclusions erode "self-evident" identities that are derived from families and local communities, and so they are part of a newly emerged globalized regime of flexible accumulation that is no longer tied to places. The exclusions are about disembodied and disembedded identities as much as they are about the increasing fragility of local communities in the face of capital disinvestment. To help understand these issues I focus, in Chapters 3 and 4, on the ways young people's bodies and sexualities are tortured and twisted by Western discourses on nature and development. In Chapters 5 and 6 I focus on the material transformations of contemporary Western society toward an incipient globalization and suggest that the changing nature of childhood is not an epiphenomenon but is, rather, central to global processes. That this is not recognized constitutes a large part of the current crisis of childhood. My sense of wonder that these connections have yet to be made is coupled with my sense of irony. To understand the relations between the exclusion of young people and the erosion of their self-evident identities on the one hand, and the material transformations of society on the other requires a fuller elaboration of the way scale is socially constructed and used to morally pound young people into place.

Scaled

> The map's scale . . . has a way of determining – all by itself – what can be seen and what can't. For instance, at small scales kids just . . . *disappear.* They get swallowed up in the worlds of their parents.
>
> (Wood 1992, 118–19)

Scale is a construct that geographers grapple with continuously and I have been wondering recently about how it might be used to address questions posed about the geographies of exclusion that permeate the lives of young people. As I argue elsewhere (Aitken 1999), scale is fashioned out of exclusionary practices that contrive a bounded and segmented social fabric. Andy Herod (1991, 84) points out that scale is not merely socially produced but it is also socially producing. At its most crass, what is produced for young people are everyday lives that are spatially circumscribed by powerful adults who, as often as not, fail to recognize the multiple ways children and teenagers shape their political identities and the scale of the day-to-day. With their heads stuck in the rarified air of power and capital accumulation, these groups – particularly male, bourgeois, white heterosexual groups – look down from an ennobled scale of privilege and prejudice.

Those who want to liberate children from privileged groups argue that these powerful adults are able to take on positions as disembodied master subjects. At its most portentous, this derives from a Cartesian objectivity that determines a view of no particular body or from no particular location, ''a view from nowhere'' (Nagel 1986). There are important geographies of scale here that help elaborate on the relations between children's experiences – their day-to-day lives, their labor and their corporeality – and larger societal issues. Iris Young calls the view from nowhere the ''scaling of bodies'' because it prescribes a politics of difference wherein subordinate groups are imprisoned in their bodies:

> In accordance with the logic of identity the scientific subject measures objects according to scales that reduce the plurality of attributes to unity. Forced to line up on calibrations that measure degrees of some general attribute, some of the particulars are devalued, defined as deviant in relations to the norm.
>
> (Young 1990b, 126)

For children the problem is particularly acute because they are expected to mature into a compliance with hegemonic norms and so they are considered ''becoming'' adults with political will rather than ''being'' children without political voice. The scale of the child's mind and body – seemingly easy to overpower – diminishes her or his will to power. In Chapters 3 and 4 I consider some precise ways Western discourse evolved in complex ways to embrace children's minds and bodies as simultaneously natural, wild, depraved, pure and innocent, but always awaiting adult tutelage and instruction. I argue that as the child's body is scaled, it is simultaneously broken, tortured and abused in a myriad ways that reflect adult sensibilities and psychoses.

Bunge (1977, 65) noted that a sophisticated understanding of scale is missing from other sciences but that ''geography recognizes that people operate at various scales simultaneously.'' In this rather narrow sense, it may be argued that geography as a discipline has a legitimate claim to study and speak for children. At his most dramatic, Bunge advocated scale changes that established a ''dictatorship of children.'' He wanted to turn the scale of power relations on their head through a revolution that installed children as the ''privileged class,'' noting that responsibility (which Bunge defined as power) would still reside with adults: ''By making children central to life, ambitions are put in scale'' (Bunge 1977, 70). Over twenty years later, through the work of social theorists like Foucault, Lefebvre, Habermas and others, we perhaps have a more sophisticated understanding of societal power relations but I still like Bunge's simple equation relating power to responsibility. In part this responsibility is simply letting children ''be all that they can be'' but an evocation of this kind – resembling closely a current advertisement for the United States Marines – is also a cautionary tale.

I will have a lot to say about adult power and responsibility on the one hand and children's rights on the other in my concluding chapter. But at this time, it is worth pointing out that I believe issues of children's public and private rights are critical in how we think about these globalizing times. Young people's rights need to be rethought in the wake of intensifying globalization and rising exclusionary identity politics not just because they have a stake in the outcome of these transformations, but because they are an integral part of them. I argue in Chapter 5 that the transformations of childhood and the transformations of the global economy are integrally linked, and in Chapter 6 I suggest that we need to rethink children's activities and behaviors in light of this link. I argue that young people's public and private rights are assaulted by moral panics over seemingly unchildlike behaviors that are in actuality deviations from outmoded norms of what constituted childhood. A specific geography of this moral assault is highlighted in Chapter 7, where I suggest that the scale interdependences of children, caregivers, their communities, environments, institutions, social structures and societies need continual and critical questioning.

Some geographers are beginning to develop a fairly coherent understanding of the relations between power and the social construction of scale (Herod 1991; Jonas 1994; Smith 1996; Aitken 1998). This book is an attempt to point towards questions that subvert scale relations that assault the lives of young people. In what ways is scale negotiated and restructured to suit and/or constrain young people and how, in turn, are places bounded by scale? How does the experience of places bounded by scale mold young people's subjectivities? Within those borders, how do class, gender, ethnic and racial struggles ascribe meanings of self on an individual as she or he contests spaces and makes places that are personally comfortable and reassuring? How easy is it to make such places and connect them with others in an increasingly fragmented world? My reasoning begins, then, with the assumption that, simultaneously at local and global scales, there are varied and contradictory geographies that both liberate and constrict the lives of young people. Despite this variation, children's voices are rarely included in the common political culture that defines public discourse and contemporary citizenship (exploitation of those voices through tokenism notwithstanding). To borrow Nancy Fraser's (1989) reconceptualization of Habermas's notion of monolithic public sphere, in a world of *multiple publics* children's voices, if heard at all, constitute the weakest of publics or are tagged onto other weak publics. And young people are *fixed* as the weakest of the weak publics because their actions are viewed as disorderly, chaotic and needing discipline (Valentine *et al.* 1998).

Fixed

The moral geographies of young people are complex and important for no other reason than they are in the process of creating future imaginaries and cultures, but there is so much more. The moral assault on young people stems in part, first, from the perception that children need disciplining and, second, from the increasingly disembedded and disembodied context within which they find themselves. One of the implications of this argument is that lived reality and day-to-day existence structures meaning and behaviors to a larger extent than a disembedded and disembodied notion of morality but it is the latter and not the former that constructs notions of childhood and adolescence. It seems reasonable to argue further that the moral value of *families* and *communities* incorporates political identities at other scales (the nation, the world) to the extent that they may be two of the most appropriate metaphors for contemporary discourses on the transformation of the sphere of reproduction. If the metaphoric use of scale is so flexible, then why are so many social objects (such as families and neighborhood groups) fixed at particular scales of material operations to the extent that their capacity to exercise power is limited (Jonas 1994, 257)?

24

Patricia Fernández Kelly (1994, 89) argues that material operations imply a translation of values into action and this is almost always shaped by the tangible milieux that encircle individuals and groups. Massey conceptualizes (1993, 61) these milieux as *power geometries* where different social groups and different individuals are placed in very distinct ways in relation to local flows and connections. Power geometries highlight difference, but they simultaneously suggest that certain actions may transcend fixed and familiar spatial grammars and so they get beyond the problematics of scale alluded to earlier.

It follows that the geographic components of these moral questions relate to how different groups of people are constructed and constituted by other groups and how these constructions distance people from each other (sometimes physically, sometimes metaphorically). I am convinced by Chris Philo's counsel that:

> moral assumptions are crucially bound up with the "social construction" of different human groupings – with deciding the character of these groupings; with laying down the codes that groups live by, particularly in their dealings with others – and this means that spatial variations in everyday moralities will inevitably be closely entangled with spatial variations in the "structure" and "functioning" of human groupings.
> (Philo 1991, 16, quoted in Holloway 1998a, 31)

What focuses the moral panics that surround the activities of children and youths are problematic social constructions of young people and the simultaneously disembodied and disembedded context of their lives. It is here that the violence occurs, a wrenching assault that is in part quixotic but also terrifyingly real when translated into the body mangling bullets that find their way onto schoolyards. It is about specific, ordinary places like suburban Jefferson County, where the myth and morality of the nuclear family was alive and well when the students at Columbine high school were massacred. The morality, of course, is part of the violence because it appropriates and encompasses the notion of a safe haven where children can be raised apart from seemingly immoral urban landscapes. The issues that are raised in the focus of Chapter 6 on the moral panic that arises from violence to and by children, then, encompass the disembeddedness and disembodiedness of young people, and an appreciation of this kind of geography may best begin with some understanding of how we come to view children and youth as categories of adult experience. The focus of Chapter 5 on the relations between globalization and the transformations of childhood is linked to the unchildlike behaviors of children elaborated upon in Chapter 6. Because unchildlike behaviors are understood most often as deviations from a globalized Western norm their practice engenders a moral panic amongst adults who no longer see the fit of "the new generations." What is required, I argue in Chapter 7, is a new interdependent and connected form of justice, and much of what I want to say about justice turns on the neo-Winnicottian view of identity, transformation and play.

The neo-Winnicottian notion of child/adult boundaries as playful, permeable and infused with meaning highlights a discussion I became part of several years ago. Concerned with some of the notions of reason and logic that emerged at the beginning of the Enlightenment project, I noted that critics at the time suggested a metaphor for human excellence abided in that adult who remains most in touch with her or his childhood (cf. Aitken 1994, 137–8). In the last decade, a significant outpouring of work on infants, children and young people suggests that an expanded metaphor is needed, one that encompasses conceptualizations of adulthood, adolescence and childhood in all their cultural, political, geographical, historical and moral complexities. It is an exciting time because some new work goes well beyond the research on child development, children's mapping and their cognitive abilities that seemed to dominate geographic research in the 1970s and 1980s. Today, a number of researchers interested in social theory and critical applications contest the knowledge bases of children and their spaces. On the agenda are issues related to children's spatial, sexual and political identities, the social construction of childhood, and how we write for and about children. Sarah Holloway and Gill Valentine (2000b, 7) argue that the contribution of geographical approaches to the new critical social studies of childhood has reached a critical mass. I want to bring partial coherency to this critical mass by arguing that although most of it does not prescribe solutions to the moral assault against young people, it none the less offers appropriate insights into difference and diversity, and the contradictory ways young people are constructed. Enduring throughout is my conviction that a focus on children's geographies articulates how places, institutions and mechanistic notions of justice teach young people how to behave and how young people resist this kind of disciplining. Although a large part of my intent with this book is to collect and synthesize work on young people by social theorists interested in geography and spatiality, by no means do I offer a comprehensive overview. Rather, I try to situate this work around a set of themes that encompass critical constructions of identities (bodies, ethnicities and sexualities) and material social transformations (local/global), and I argue that some of these contradictions find accommodation in a reworking of notions of justice, privacy and play.

2

FROM THE GROUND UP

Natural developments

An exchange in the journal *Area* in the early 1990s suggested that geographic research on children is characterized by diverse ways of knowing that lack intellectual coherence and direction (cf. James 1990; Sibley 1991; Winchester 1991). At the time I argued that this was not necessarily so, and that research on children's geographies had a fairly robust intellectual history spanning at least two decades (Aitken 1994). For the most part this intellectual trajectory followed a scientific discourse with interests ranging from children's spatial knowledge acquisition or wayfinding abilities to their understanding of larger geographic concepts and what Blaut (1991) refers to as the macro-environment. The child's world was constructed as a Cartesian space that opened up with increased knowledge and development. This space was a container for the child's activities. To a lesser extent, geographers followed humanistic interests and developed ethnographies that spoke to children's experiences of place and nature. From this discourse, the child was conceived as a monadic being and external social, cultural and moral contexts, although important, were secondary to understanding the essence of child–environment relations. Intellectual coherence was formed by highlighting questions that related to how children's perceptions, attitudes and opportunities were structured within environments. Again, space was a container for children's activities.

Until very recently, and like most other social scientists, geographers worked within commonly held assumptions about children, often without attending to the moral, cultural and political contexts of those assumptions. This chapter outlines these assumptions and attempts to trace the ways they resonate with larger intellectual transformations. I begin with a consideration of how we came to conceptualize childhood as innocent, naïve and closer to nature, and how this perspective evolved in complicated ways through scientific and biological discourse to our understanding of children as *tabula rasae* upon which adult codes may be written. These codes are writ large in the developmental theories of the nineteenth and twentieth centuries that paved the way for a child-centered pedagogy that dominates to this day. My argument throughout is that these discourses do violence to the actuality of young people's experiences and this violence is brought to light only recently by critical and post-structural theories of identity that account for the moral, cultural and political contexts of lived experience.

The last decade witnessed a turning point in our understanding about young people because there was recognition of unsubstantiated knowledge about space and about children. It was recognized that research within paradigms that assumed space as a mere container or stage for children's activities was perched on precarious and unsteady intellectual foundations. There was also a realization that talking about childhood and adolescence as categories of experience was quite different from the study of young people's daily lives. The former is imbued with meanings that reflect larger societal conceptions about the nature of experience and, in particular, what it is to be not-an-adult. The latter is focused and directed by a concern over the everyday contexts of children's lived experiences as they are socially and spatially structured, and, ultimately, by questions about how cultures are reproduced through children. Of particular importance with the turn to critical and post-structural perspectives in geography and other disciplines, is that theories, categories and methods assumed unassailable only a decade ago are now questioned and discarded. What I want to do with this chapter is trace some of the ways geographers have come to study children and to know childhood with a specific focus on intellectual and scientific underpinnings. In a sense I am, once more, questioning and discarding a swathe of perspectives that suggests childhood is a specific and knowable state of being and that child development follows certain prescribed stages and processes but I am also laying the foundations for discussion of the embodiment of childhood. The notion of embodied children embedded in local/global contexts is a fulcrum upon which the main themes of the book – the nature of childhood, bodies, sexuality, justice, moral authority, privacy and play – fitfully swing. My intent in this chapter is to outline the ways "nature" and "the natural" prescribes notions of childhood and then to turn questions of natural development on their head in the next chapter by focusing on the ways children's "natural" experiences are disembodied by twentieth-century developmental theories.

Scientific and biological discourses

The natural (sometimes known as biological or nativist) perspective suggests that children see the world in ways similar to those of adults and development is simply a question of growth. Questions of what constitutes growth are usually dropped into the miasma of self-evident truths because most children grow and at some point become adults. But what is meant by growth and how does it take place? Can emotional and intellectual growth be accounted for in the same way as

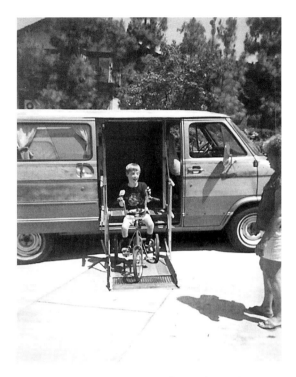

Figure 2.1 Don and his tricycle. Don has cerebral palsy. When we met he wanted to show me how he was able to ride his tricycle up into his Dad's van. This was one of Don's many recent accomplishments. According to school and health officials, Don is developmentally three years behind his "able-bodied" and "neurologically normal" peers

Source: Joan Isaacson

physical growth? How do we know when these processes are complete and the child is an adolescent or an adult? What parts of growing up are "natural," and what parts are not? If a child has cerebral palsy or some other neuro-motor or physical impairment to their "normal" growth, are they then abnormal? Are unnatural developments to be blamed on pathology, environmental mishaps or inadequate parenting? The biological perspective may answer these questions in fairly precise scientific language that can, for example, establish "normality" from a statistical bell-curve but it also sidesteps the moral implications of labeling those on the tail of the curve as "abnormal." It is not within the purview of the biological sciences to probe the possible social, cultural or institutional causes of unnatural developments.

Much biological science, in its attempt to pin down developmental patterns and processes, frequently sidesteps the notion that scientific responses not only influence, but also embody social and cultural shifts. Thomas Laqueur (1990)

29

reveals shifts in the biological sciences' historical constructions of men, women and sexuality from Plato to Freud. The physical players in Laqueur's story are the human sexual organs, food, blood, semen, eggs and sperm, but the plot is created and narrated by scientists, politicians, literary theorists and intellectuals of all persuasions. His point is that biological sex is not just a social construction, but that almost everything that anyone wants to say about sex is explicable only within the context of battles over identity politics. It is not difficult to point out similar battles about growth and child development. For example, when teething was redefined by science as a normal, natural event in eighteenth-century medical writings, its status as a disease was rejected. Also at this time, Rousseau's criticism of the widely practiced use of wet-nurses as unnatural heralded what has been called a cult of breast-feeding (Bloch 1974). Then, at the beginning of the twentieth century, breast-feeding again went into decline in North America and Europe as mothers were suspected of transmitting diseases to children through their milk but also, at the same time, special ''formula'' milk was produced for mass consumption.[1] Testing their new-found cultural authority, pediatricians concocted and marketed an assortment of infant formulas and special bottles designed to replace breast-feeding (Gollaher 2000, 100). The important point that these examples elaborate was not that people suddenly changed their child-rearing habits, but that a group of middle-class professionals and scientists rethought an aspect of childhood as part of their approach to nature. And market pressures reinforce this ''nature.'' A related theme that doggedly follows this chapter and the next two is that discussions about children are suffused with moral assumptions about the body and even when couched in the language of science, these discussions are always a way of speaking about other concerns.

Our corporeality – the very ways we think about our being – suggests the importance of links between growth, development, the body and sexuality but a large part of scientific knowledge attempts to understand these links by parsing out ''the nature of childhood'' into linearly related but none the less distinct compartments. Post-structural theories suggest non-linear and complexly woven links that are not so easily explicated. They suggest further political, social, economic, cultural and moral values that attend the parsing of young people and their development. The language of natural science hides these values, lending credence to a specific way of knowing nature and children with a long history in Western thought.

The natural child

It is reasonable to begin with an expectation of controversy concerning what a child brings with them into the world. Historian Ludmilla Jordanova argues that Western science over the last three centuries tends to refer to children in terms of nature and that this practice continues as an imprisoning discourse: ''This long-standing analogy is reinforced by our lively biological sense of the processes of procreation, a fresh consciousness of children – at least when babies – as wonders of nature'' (Jordanova 1989, 6). At its base, the conflation of children with natural imagery (pure, innocent, plantlike, tender) is related to thinking of children as asocial or presocial. That said, the polyvalency of nature suggests an unbridled complexity in the ways children and childhood are known. The state of childhood is sometimes seen as one of physical beauty. It is also seen as a state of innocence and purity wherein children are thought to possess wisdom that is no longer accessible to adults. From this perspective, children may also be feared for their seeming animal-like instincts. From a less aesthetic perspective, children are sometimes categorized alongside animals or plants as natural objects available for scientific and medical research. Each of these views is steeped not only in moral values but also in particular class/gender struggles that speak to tensions arising from particular cultural histories and geographies.

It is generally accepted that children's innocence was first established in the mid-eighteenth century by Jean-Jacques Rousseau who argued in *Émile* (1962, originally published in 1762) that all things begin good and it is only through the social world and ''man's meddling'' that they become evil (Jenks 1996; James *et al.* 1998). The story of *Émile* relates the growth of a boy from infancy to manhood and how his development is directed wisely by a tutor. Given its general acceptance as the harbinger of a particular view on childhood, it is important to note that the form of education described in the book was inaccessible at the time to all but the most wealthy. Of course, this kind of learning required women to be at home and always available for their children.[2] Education of the kind Émile aspired to was solely for men:

> The art of thinking is not alien to women, but they only need a nodding acquaintance with logic and metaphysics. Sophie [Émile's fiancee] forms some idea of everything, but most of what she learns is soon forgotten. She makes best progress in matters of conduct and taste.
>
> (Rousseau 1962, 157)

If these sexist and classist assumptions are bracketed (but not dismissed), what can be made of Rousseau's enlightened concern about the state of childhood and its relations to nature? He begins by highlighting that everything from God is good but degenerates once it gets into the hands of man: ''Not content to leave anything as nature has made it, he must needs shape man himself to his notions as he does the trees in his garden'' (Rousseau 1962, 11). There is, for Rousseau, a natural, pre-social moral goodness in children from which adults benefit and they need to cherish this so that it may form the basis of tomorrow's society. At one level, this instills a sense of children as innocent and untarnished purveyors of good but, at another level, it suggests the question of a child's particularity. Rousseau established the intrinsic value of the term childhood in terms of understanding the child as a child. And so, as social psychologist Allison James and her colleagues argue, contemporary concern about children as individuals stems from Rousseau's rhetoric: "For the first time in history, he made a large group of people believe that childhood was worth the attention of intelligent adults, encouraging an interest in the process of growing up rather than just the product" (Robertson 1976, 407, cited in James *et al.* 1998, 14).

Rousseau's ideology becomes the kernel of ideas that question notions of children as little adults or incomplete adults, and introduces them instead as beings with rights, but not necessarily obligations, beyond their own innocence. Nature, after all, is not obligated to protect or nourish humankind. The responsibilities befall adults in the form of protecting children and bringing them up in a manner that protects their best qualities from the violence and ugliness of contemporary society. As one seventeeth-century commentator put it: "Familiarize oneself with one's children. . . . They are young plants which need tending and watering frequently" (Goussault 1693, quoted in Ariès 1962, 132). Importantly, then, Rousseau not only established the particularity of children but also their moral innocence as things close to nature.

If children are revered as innocents they are equally naturalized as instinctual wild animals that are amoral and even evil. Jenks (1996) argues that the idea of children as unruly and unsocialized "little devils" emerged prior to the notion of children as "innocent" or "angelic." Part of his argument comes from sixteenth- and seventeenth-century Puritanical socialization practices such as "not sparing the rod to save the child," which at a time of high infant mortality rates advocated harsh disciplining so that children might be ensured heavenly salvation (Richardson 1993). Another suggestion that is linked to Puritanism is that the will of children needed to be tamed in much the same manner that the spirit of feral horses is broken so that they lose their naturalness and become compliant and domesticated.

A century after Rousseau, and from a radically different philosophical position from either his or that of the Puritans, the transcendentalist appreciation of nature espoused by Thoreau, Emerson and other American romantics elaborated the instinctual wildness of children onto nineteenth-century thinking:

> No human being, past the thoughtless age of boyhood, will wantonly murder any creature which holds its life by the same tenure as he does. The hare in its extremity cries like a child. I warn you, mothers, that my sympathies do not always make the usual *philanthropic* distinctions.
>
> (Thoreau 1993, 177)

Many credit Henry David Thoreau (1817–62) as the originator of American ecology in that he emphasized the interrelatedness of the natural world. Like many of his contemporaries, Thoreau regarded nature as opposite, and an antidote to, the excesses of early industrialization but he also championed the notion that people are indelibly linked to nature. Indeed, Thoreau's approach to the natural world is very much a response to the early forces of globalization and his famous studies around Walden Pond attempted to reconstruct a pre-colonial American landscape. Ralph Waldo Emerson was a neighbor of Thoreau and, as a member of the romantic school, mapped transcendental spiritualism onto Thoreau's pragmatics. Emerson's urban tastes (he was raised in Boston) resulted in an urban/ rural tension that he resolved in favor of the country: ''I wish to have rural strength and religion for my children and I wish city facility and polish. I find with chagrin that I cannot have both'' (cited in White and White 1962, 27). This resolution, argue White and White (1962), is a direct result of changes in a rapidly industrializing Boston during Emerson's lifetime with many new railroad lines penetrating the spacious gardens and open hillsides that he knew as a boy. As with Thoreau, globalization and the changing moral environment of cities repelled Emerson. The importance of global and spatial transformations is linked for these writers to changes in how children should be viewed. The transcendental philosophy of Emerson and Thoreau suggested that a person could rise above nature and the limitations of the body to a point where a spiritual life replaces the primitive and savage: ''We are conscious of an animal in us, which awakens in proportion as our higher nature slumbers'' (Thoreau 1993, 182). For Thoreau, children were a central locus for the natural and the ''animal in us.'' Education, spirituality and a grasp of ''higher laws'' took adults to a moral plain that transcended the negativity of urban industrial society, initiating a state of grace and commune with nature.

Assumptions about the nature of children laid down by Rousseau and Emerson span two centuries and two continents, but they comprise a foundation that influences a large swathe of contemporary Western thought and social science practice. Children are embodied with a wild nature and the question for contemporary times falls between whether it should be encouraged or disciplined.

Wild thing

Gill Valentine (1996) argues that although what it means to be a child varies over time and space, since Rousseau's romanticization of early innocence the Western constructions of childhood have oscillated between representing children as vulnerable ''angels'' or wild ''devils.'' These mythic constructions are represented in the way geographers and other social scientists have studied children but they are also a large part of popular culture today. I want to use two brief examples that ascribe a "wild nature" to children, one that appears rather banal in contrast to one that elaborates an insidious racist subtext. I argue that both are quite startling in their implications of children as wild things.

In a popular best-selling book entitled *The Geography of Childhood* (1994), Gary Paul Nabhan and Stephen Trimble use a series of anecdotes and essays to highlight why children need wild places. They begin the book by proclaiming that "the geography and natural history of childhood begins in family, at home, whether that home is in a remote place or in a city," suggesting an interesting conflation between natural history and childhood with an equally interesting foundation in the home/family that I pick up in the next section. The basic argument throughout *The Geography of Childhood* is that "the earth" and "wild places" are more than therapeutic, they provide answers to some of life's most tenacious problems. In a chapter that discusses differing gender roles and the power (often violent) of boys over girls, men over women, Trimble argues rape undermines trust and that "once again, our best hope lies in helping to teach our children somehow to combine masculine and feminine, integrating sensitivity and strength, turning to the earth as a setting and source for both" (1994, 63). Although the discussion is otherwise sensitive to the oppression of women, it misses the larger story of how nature is constructed as feminine and how that contributes to the oppression. It misses the construction of nature (and women) as exploited and dominated by white, middle-class men in Western society (Warren 1987, 1990; Davis 1988). It also sidesteps ecofeminist attempts to replace the traditional deceit that "woman is to nature as man is to culture" with views that encompass the values of nurturing and emotion without perpetuating the dualistic myths from which they derive (Rosser 1991; Plumwood 1992). Indeed, aspects of the book embrace these dualisms by suggesting (seemingly from a gender neutral perspective) that such antinomies as culture/reason/domestication and nature/emotion/wildness are separate and contextualize children's play in different ways. For example, in the penultimate chapter of their book, Nabhan elaborates on the larger scientific discourse from which his collective work with Trimble derives. He argues (1994, 155–6) that our "wild side" is related to that part of the brain in which our most primitive behavior resides and that this is precisely "the ancient animal which some of us care not to acknowledge as part of our being." In his view, children should be encouraged to know wild things and to greet this "other" with a sense of respect and kinship.

Children getting in touch with their inner wildness by handling and getting to know lizards, snakes and frogs is one thing, but something different happens when a wild nature is ascribed to adolescents. A poignant case of this kind focuses on the attack of a woman in New York's Central Park by a group of Latino and African-American youths. In an article that highlights the complicity between writing in the social sciences and constructions of a nature that is to be dominated and controlled, Cindi Katz and Andrew Kirby (1991, 265) raise the Central Park case as the example that turned the "aestheticization of natural environments" in urban areas on its head with the notion that these too are "wild" places. Urban parks, they note, are intended to be restorative preserves of "nature," but on the night of April 18, 1989, when the 28-year-old white female jogger was brutally beaten and raped, authorities and the popular press immediately took up the notion of "wilding." Within hours of the attack, the police rounded up two dozen teens from nearby Harlem. Katz and Kirby document how the authorities intimidated and coerced the youths to extract confessions from several of them (some of the methods used infringed on their rights as minors). They note that the rape of the "jogger" (read middle class and white) and the trials that followed dramatized and increased a series of vicious class, race and gender divisions in New York. These divisions enhanced speculations on the reason for the attack and why this particular case was dramatized. Several contemporaneous rapes and murders of minority women received very little press.

Katz and Kirby argue that at least part of the case's popular appeal stemmed from the police attributing the attack to "wilding," and the authorities' claim that this was a term used by one of the suspects:

> Although it appeared to be an offhand remark or a misheard term, the press announced the apparently new phenomenon in banner headlines and initiated their own investigation. "Wilding," according to press and police reports, took place when male adolescents grouped together on a frenzied rampage, attacking people and property indiscriminately. Soon the press referred to the suspects as "animals," a "wolfpack" out of control in the park, and reports of their attacks and confessions grew even more lurid.
>
> (Katz and Kirby 1991, 267)

Combining social imaginaries that derive from the racism and recapitulationism that inspired Joseph Conrad's (1946) stories "Youth" and "Heart of Darkness" and William Golding's *Lord of the Flies* (1954) respectively, the notion of "wilding" fitfully rests with white middle-class fears of the uncontrollable natures that reside close to home. And, as I articulate more fully in Chapter 4, which deals specifically with race and space, *place* is also important in the construction of the wild. As an aestheticized place, Central Park (a natural environment in one of the most densely built cities in the world) was built and romanticized to represent a garden, a tranquil oasis that afforded respite from urban pressures. By the late 1980s it has seemingly evolved into a wilderness (which has etymological roots in the Welsh and Old English words, *gwylt* and *deoren*, meaning wild beast), no longer tamed and compliant. The events articulated by the authorities

35

and the police and ritualized as "wilding" are interpreted by Katz and Kirby as a subversion of the social mechanisms of control. As an expression of white middle-class fears, the notion of "wilding" conflates some intense feelings that encompass minority youth and the construction of nature.

I raise these two examples – one of popular child/nature writing from two fairly well-known naturalists and the other about the construction of young people as wild and barbaric – not to denigrate the very important work on deconstructing the social and political relations between humans and animals (cf. Anderson 1997; Wolch and Emel 1998), but to highlight popular contemporary notions of young people as naturally predisposed to wilderness and wildness. Wolch and Emel (1998) collect together a diverse set of essays to illustrate the significant role that animals play in the construction of the social environment. As one reviewer of the volume asserts, the problematic nature–culture divide loses its integrity when the space and place of animals and nature within the social environment is contested (Lulka 2000). But the two examples described here are about conflation and reification rather than contestation. With focus on the unequal social relations within and between species, it is important to attend to problematic, and popular, metaphorical exchanges about the natural dispositions of young people. Two hundred years after the publication of *Émile*, young people are still thought to be naturally closer to nature with little thought to how childhood is constructed as closer to nature. Part of that construction, of course, comes from powerful associations with other media besides journalistic texts that point to the ways that children and nature are jointly commodified and consumed.

Consuming nature

In their work on child/environment relations, Bunting and Cousins (1985) note that early predispositions towards natural environments may come, at least in part, from the media. Many of the very first images presented to children teach them about a wider world that is heavily biased towards mythic representations of nature. The narrative of enduring children's books such as those written by Beatrix Potter, Thornton Burgess, Enid Blyton and A.A. Milne, and the equally successful, and "timeless," films from the Walt Disney and Dreamworld studios are based almost entirely on pastoral contexts that glorify nature and animals: "In these stories of and about childhood, children's presence in the country is naturalized: children are portrayed playing outdoors, with companions, beyond the surveillance of adults, blessed through their proximity to and interaction with nature" (Holloway and Valentine 2000b, 17).

With the exception of *Sesame Street* (notably, the most successful North American TV program for pre-schoolers), contemporary television materials aimed at young children seldom focus on urban environments per se and when they do there is none of the romantic glorification that characterizes comparable types of material portraying the natural environment. Most children's programs and cartoons not only venerate animals and nature in pastoral settings but are couched within very specific patriarchal discourses. A film such as *Bambi* (1962), for example, sets the forest animals as heroes against ''man'' (the hunters), but also establishes explicit patriarchal relations between Bambi's mother (who is in charge of the children) and his father (who only ever appears in times of trouble to save the day). At the end of the movie, Bambi's mate gives birth to two fawns. Where is Bambi? He is on top of a ridge with his father, above and removed from the birthplace. In the next scene which closes the movie, Bambi's father walks into the sunset leaving his son as the new patriarch and protector of the forest below. Not only is nature venerated, but so too are patriarchal relations. In *The Jungle Book* (1965), Mowgli successfully resists the temptations of civilization until a young woman coerces him with a song about her hopes to establish a family with a ''father out hunting in the jungle a mother at home cooking by the fire.'' The message is clear: these kinds of gender relations are the ''bear necessities'' of life. Dominant sexist and racist discourses are also found in popular cartoons which espouse an explicit environmental theme. Bill Kroyer's *Fern Gulley* (1992), for example, represents the Amazonian rain forest but there are no Latin American characters amongst the fairies and humans, and all the music is Western. The popular *The Land Before Time* series from Universal Cartoon Studios embodies implicit environmental themes as does Disney's *Dinosaur* (2000). The narrative of *Dinosaur* draws heavily from the original film in the *The Land Before Time* series suggesting, in particular, a timeless search for the sanctity of a lush hidden valley free from the threat of predators (Carnosaurs and other ''sharp teeth''), led by a juvenile male with a supporting cast of emotionally less well-balanced creatures. Don Bluth's science fiction cartoon, *Titan A.E.* (after earth) (2000), targets teenage boys, and the personality traits of the hero bear a strong resemblance to those of the juvenile male in *Dinosaur*. Although the bulk of the action takes place in the machine-like future worlds of derelict space-ships, the films ends with the hero ''getting'' the girl and actually creating (using God-like science invented by his father) a spectacular new natural world with resplendent waterfalls and oceans. Not surprisingly, the new world is constituted by the design of the father – who leaves the boy as a child and dies protecting his own vision for humanity. These cartoon stories elaborate potentially powerful hierarchical codes, suggesting problematic myths to live by.

Thinly veiled patriarchal discourses are not limited to the fantasy of cartoons. Leo Zonn and I (1994) show how an environmentally ''focused'' film for children, *Storm Boy* (1976) helps to perpetuate and bolster a series of myths of male dominance in Australian culture. The film was supported by the Australian

Film Corporation and was accompanied by a large marketing campaign that included a twenty-minute documentary on the making of the film and a "Storm Boy Picture Book" for pre-schoolers and the "Pic-a-Pak Study Guide" for schools. That the film holds a significant debt to environmental conservation subtly subverts the familiar ecofeminist theme of "man supporting culture" and "woman supporting nature." Women, although largely absent in *Storm Boy*, represent urban, civil society – a society from which the main adult male characters (one aboriginal and one white) find sanctuary in a natural preserve. The parallel with Mowgli's entrapment by the young women is quite evident. It is within the environment of the nature preserve (untamed jungle) that "Storm Boy" (the son of the white male) grows up and it is with the tension of urban–society–female versus nature–freedom–male that he and his father are forced to come to terms. The individuals of the story clearly exhibit features of the purported Australian character (anti-society, freedom, mateship, up against an indomitable physical environment). In fact, the young Australian viewer may have been presented a primer on the nature of the national psyche. The relation between Storm Boy and his pelican, which is of foremost interest to many children, is clearly that of mateship and anti-authoritarianism. Best of friends, they seemingly are bonded by the environment and together they oppose the world of those people who can harm them; from a woman teacher who wants the boy back in school to the hunters and others who harm the environment. But it is the pelican that ultimately, and subtly, reinforces a patriarchal societal ethic. The pelican returns to Storm Boy after being freed into the wild, suggesting that it is all right to shun nature and freedom, ostensibly, for friendship, but also for the constraints of community. It is only through the pelican's death (sacrifice) that Storm Boy's father resigns himself to a move to a local town and the values of the school. And it is through the pelican's heroic efforts in saving a doomed yacht's crew that the father receives a reward sufficient to purchase a petrol station (and a new life) in town. Will he marry the school teacher to provide the best kind of happy ending?

Jack Zipes (1997, 91) points out that most producers of films, and particularly Disney, are interested in "hooking" children as consumers not because they believe in the importance of child–nature relations or that their films have artistic merit and could contribute to children's cultural development, but because they want to control children's aesthetic interests and consumer tastes. This is a far cry from genetic predispositions that are thinly veiled in the work of Nabham and Trimble. It is evident that there is also a strong patriarchal bias to these representations of child–nature relations. To accept that Rousseau's *Émile* set the foundations of Western thought of childhood as particular and innocent is also to accept certain class and gender biases. As I pointed out earlier, the education of Émile was not one that could be readily imitated except by people of rank whose children were brought up by tutors in private schoolrooms.

I raise these issues to suggest that the complex relations between children and nature are not just about childhood innocence. They are also about the weaving together of mythic representations of children and representations of nature that derive from cultural geographies that reflect gender and power relations, and that these relations have a complex and long history, and that this history manifests itself in today's geographies. Valentine (1997b) explores how nature and rural areas are idealized as places in which to grow up. In particular, she considers how parents mobilize popular notions of the rural idyll in accounts of opportunities afforded their children in a small English village. What is interesting about this work is the way that parents living in the village raise issues of safety that contradict popular contemporary constructions of the rural as a safe and harmonious place, ideally suited for the raising of children. To offset these contradictions, some parents found solace in the notion of the village providing an appropriate kind of community within which to raise a child if the surrounding countryside did not. Valentine uses the study to argue that it is important to recognize that places and children have multiple meanings and identities. Different understandings of nature and ''the rural'' co-exist and children are contextualized in these quirky environments in different ways. Clearly, understandings of children and childhood cannot be universalized beyond the cultural geographies and histories within and through which child and adult lives are contextualized. Although the local contexts are important, what is interesting about Valentine's work is that it raises the importance of other myths that also guide actions and values. In complex ways, the myth of nature and the rural idyll are conflated with seeking appropriate places to raise children and embedding families in particular kinds of community. I argue elsewhere that there is a ''structural fragility'' to these mythic constructions that make it very difficult for most contemporary families to achieve even a semblance of what these myths supposedly offer because of the disciplining effects of contemporary spatialities (Aitken 1998). Put another way, in that the search presupposes that the good life exists and is attainable in specific places, the myths endure as a source of tension and frustration for many families. In some ways, nature, family and community are seemingly universal forms of knowledge that say very little about a child's growth and development but in other ways they are mythic and problematic foundations for that growth and development. In large part, the growth and development of children is more often than not the outcome of negotiating and enduring these mythologies of social reproduction.

Viewing children as instinctive animals or innocent progeny of nature not only universalizes some very complex local geographies and gives them a form of otherness, but when conflated with mythic family and community forms it also establishes certain kinds of priorities on how children should be enculturated and socialized. This suggests different foundations of childhood than those that are primitive and ideal, and it draws me into a discussion of another basis for socialization that begins with the premise that the beginning child is a *tabula rasa* upon which society etches the nature of childhood. Up until now, epistemo-logically, when considered as "real," nature and children are a privileged domain with a specific and pure form of knowledge that is not necessarily affected by socialization. It is clear, however, that this primitive, ideal reality is extremely problematic because it ascribes a specific form of knowledge to children. The alternative view, that there is no specific form of knowledge attributable to child-hood, is perhaps best represented by the empiricism of John Locke and the twentieth-century developmental theories that drew heavily from his work. In what follows, I begin with Locke and use his work to elaborate on the onto-logical basis of child development theories and suggest that this work is still tied heavily to mythic notions of child/nature relations. I argue that the influence of Rousseau's naturalism and Locke's empiricism blend problematically in the twentieth century with Darwinian and Freudian theory on progress and sexuality respectively to create a monolithic torpor wherein understandings of childhood and child development are disembodied and removed from a practical exegesis of day-to-day life.

The developing child

Empiricist views suggest that a child is born into a world of complete chaos and that skills and knowledge develop only with the right environment. From this perspective, no knowledge is innate but rather derives from experience, and this derivation proceeds slowly. John Locke is considered by many to be the pro-genitor of empiricist and analytic traditions in philosophy. Although he had very little to say specifically about children and childhood, David Archard (1993) argues that one of his main philosophical treatises – *Some Thoughts Concerning Education* (1693) – might be considered along with Rousseau's *Émile* as a harbinger of child-centered education and also a root of child developmental theory.

A group of seventeeth-century academics at Cambridge University known as the Platonists took the notion of childhood innocence a step beyond Rousseau by suggesting that the fundamental nature of children derived from innate good-ness rather than innocence. Rousseau's notion of innocence suggested merely a form of neutrality from which children could be either angelic or wild, but the Cambridge Platonists advocated childhood as a time suffused with forms of sym-pathy and benevolence upon which adult sociability is based (Valentine 1996, 584). Locke counters this by suggesting that although children may be "becoming adults'' and citizens-in-the-making, they are none the less *tabula rasae* waiting to be filled with experience and reason. For Locke, both knowledge and rationality are obtained incrementally. He sidesteps issues of children having any innate pre-dispositions to certain forms of knowledge – either wild or angelic – by arguing that the only inborn characteristics of children are those processes through which the mind can reason. In this regard, children are no different from adults with the exception that adults have more time to reflect and more to reflect upon. Children are not rational although they are born with the capacity to learn reason. They are on their way to becoming adults and education is a fundamental process through which adulthood is achieved (Archard 1993, 3). A virtuous person is produced through appropriate environmental contexts, and the achievement of a reasoned mind is accomplished through relevant and appropriate forms of education. No form of knowledge is innate or native, but the capacity to reason is both innate and develops through education, maturation and experience.

Darwin and natural child development

Locke's ideas set the stage for childhood as a serious object of scientific investiga-tion beginning in the nineteenth century, but ideological connections with nature do not necessarily diminish with the empiricism that distinguishes his work and that of his followers. Indeed, if anything, the connection between childhood and nature is reinforced at this time by the work of Charles Darwin. Darwin's complex influence on geography through the first half of the twentieth century is well documented by David Livingstone (1992), but his contribution to con-temporary understandings of children's geographies is appreciated with a more nuanced focus on how he conceived of the human mind. As part of his work, Darwin (1887) made a study of his own son: ''A biographical sketch of an infant.'' The basis of this study, of studying and observing a human child under the same terms and conditions as other ''species,'' is important for in this act fomented the modern idea of ''natural child development'' (Walkerdine 1984, 170). Darwin sought the evolutionary link between humans and lower animals through empirical science, but he also believed that his evolutionary hypothesis was traceable in the child's ascent to adulthood. Indeed, Livingstone (1992, 182) points out that reading an anthropomorphic element into the workings of natural selection ''was a temptation that Darwin found hard to resist.'' That the development of the species may be, at least metaphorically, read in the devel-opment of an individual's mind suggests a belief that children develop towards adulthood out of animality (Archard 1993, 32–3). This notion developed into the theory of recapitulationism which states that an individual lifetime reproduces the patterns and stages of development in the species in a vastly scaled-down time frame (Davis and Wallbridge 1981).

Livingstone (1992, 187) argues that the most influential extensions of Darwin's work on the social sciences came from a rejuvenation of the earlier evolutionary doctrines of Jean Baptiste Lamarck. Neo-Lamarckianism suggested that the response to environmental influence can be inherited and transmitted through the action of natural selection. It posited two connections between evolution and society. First, the doctrine of the inheritance of acquired characteristics held that the qualities gained by an organism in its life experience would be passed on and, second, the directive force of organic variation was attributable to will, habit or environment. Neo-Lamarckianism provided an account of natural variations based upon environmental differences and the force of will and reason in equal amounts. In geography, both tenets of neo-Lamarckianism provided a set of principles upon which environmental determinism and possibilism could comfortably reside through the 1940s.

The notion that environmental responses are based on genetic memories was picked up in Jay Appleton's (1975) habitat theory. This theory was an influential basis for scientific and humanistic research on person–environment relations through the 1980s. It influenced a generation of scholars who were searching for a theoretical explanation for why certain environments seemed more prefer-able than others. Appleton suggested that people advanced beyond other animals on the African savanna through developing abilities to know the location of predators by standing upright, climbing trees or appraising the larger environment from a hilltop. Understanding the values of these places with ''prospect'' along with others that offered ''refuge'' (e.g. caves, treetops) gave early people an evolutionary advantage based upon habitat. Appleton contended that these early places of survival are now etched onto our genes to the extent that we derive aesthetic pleasure from tree- and mountain-top views, and a sense of comfort from the telluric spaces of caves and deep canyons. These feelings are universal and are most evident in childhood where they have not been eroded by accultura-tion and other social pressures.[3]

For the nascent field of developmental psychology at the turn of the century, neo-Lamarckianism and recapitulationism suggested the importance of how children come to know environmental settings and how reason develops to enable adaptation to those settings. Perhaps the greatest advocate of these views was Granville Stanley Hall, professor of psychology at Johns Hopkins University and Clark University. According to Hall (1904, 1909), genetic ''memories'' about the emergence of humans from the primal muds to the so-called tribal phase of primitive peoples were recapitulated during childhood. This recapitula-tion stimulated instinctual desires to re-enact phases of human evolution. The desire of young children to play in sand and mud presupposed a genetically induced feel for the primal ooze and then, later, the desire to form gangs was a response to the instinct for tribal life.

In an influential set of studies in the early 1980s, Trudi Bunting and her colleagues worked with over 1,000 children in Ontario, Canada seeking to understand children's dispositions to certain kinds of environments (Bunting 1983, 1986; Bunting and Cousins 1983, 1985; Bunting and Semple 1983). Using rating scales designed by environmental psychologist George McKechnie (1974, 1977) and drawing on Erik Erikson's (1969) notion of how environmental mastery and the self develop, they assessed children's preference for different environmental settings. On the whole, they found that children were predisposed primarily towards natural or pastoral settings. The rating scales measured more than simple scenic preferences, including also intellectual and aesthetic enjoyment of nature. It is important to note this because it is possible to argue that Bunting's scales focus on particular views of what constitutes a child. Indeed, Appleton's work is cited by Bunting and Cousins (1985, 731) as influential in contemporary understandings of how environmental cues are determined by human evolutionary theory: ''It is argued that the genotypical response is more marked in children's environmental preferences because children's preference is less likely to be influenced by either sociocultural learning or direct environmental experience.'' Bunting (1986) argues further that children's changing attitudes with age reflect a growing sense of control and challenge and she assumes that this relates to developmental theory.

The neo-Lamarckian tension between inheritance and instinct, free will and reasoning, and environmental influences contextualized the course of scholarly thought in child development throughout the twentieth century. Hall argued that children must successfully re-enact each evolutionary stage to develop ''correctly'' and, in later work, he and his students suggested that these stages could be tied to particular ages. Although most later developmental theorists shied away from recapitulationism, the notion of prescribing a correct form of development and tying it to specific ages survived through most of the last century.

Staging development

The most influential developmental theorist of the twentieth century, Jean Piaget, was persuaded by the Darwinian notion that children enter the world with a genetically transmitted nature and the neo-Lamarckian notion that this nature enables them to adapt to their environments. He also borrowed from Locke's belief that a child's first experiences are exclusively sensory and that although some capacity to reason is inborn, for the most part it requires education, maturation and experience. According to Piaget (1952), reasoning about the environment is acquired in the normal course of human development. Where he differs from Locke is in his dismissal of the idea that a child's ability to generate knowledge incrementally unfolds with experience. Piaget dismissed the idea of the quantitative accretion of environmental knowledge in favor of the idea of qualitatively distinct stages of development. As a structuralist, then, he believed that there exist specific stages of cognitive development; but there is another aspect of Piaget's work that bears noting. As a constructivist, he believed that what we take as real is a construction of thought (Piaget 1971). Like Locke, Piaget believed that ways of knowing are less likely to be inherited and more likely to form through complex interactions between maturation and socialization, between children and their environment. By adapting to change, intelligence develops for Piaget through the child's active participation in the environment.

Erica Burman (1994) explores how the motivation and context of Piaget's work enabled and guided the incorporation of his theories into the dominant empiricist culture, following Hall's lead to establish a biologistically based model of intellectual development. Burman also shows how wide-ranging is Piaget's influence in many social sciences. Certainly, his influence on research in children's geographic thinking, education and child psychology is particularly significant (see Downs *et al.* 1988, 1990). Indeed, his theories form the basis of the cognitive approach to psychology and today they virtually structure the discipline of developmental psychology.

Perhaps most appealing about Piaget's theories was their founding in, and mutual support of, an analytic and empirical philosophy of science. There were no empirical tests to validate the work of Hall and Appleton, but Piaget's theories stood up to fifty years of experimental testing in a wide array of disciplines. Piaget conjoined a new set of theories with new forms of empiricism.

In his methodological work, Piaget was quite critical of the empirical foundations of psychology. Through a startling battery of novel experiments and innovative laboratory tests that included toys, ball games, hide-and-seek, sketch maps, geometric manipulations, mapping exercises and tests for motor speed and agility, Piaget and his students amassed empirical evidence to suggest that there are four major stages of intellectual growth: sensorimotor, pre-operational, concrete operational and formal operational (Piaget 1952, 1954; Piaget and Inhelder 1956). Piagetian theory prescribes for most children a linear, normal and natural form of development where each stage is quantitatively and qualitatively different from that which precedes it. These stages and their relationship to spatial understanding have been described in other geography texts to the extent that only a brief overview is needed here (cf. Hart and Moore 1973; Moore 1976; Hart 1979; Walmsley 1988; Matthews 1992; Aitken 1994). Piaget asserted that infants have only limited representations of their world, and it is only towards the end of the sensorimotor stage that any intelligence begins to form. In terms of spatial orientation, children are exclusively egocentric wherein all things are located relative to (or even as part of) the child. There is little or no division of subject and self in the Freudian/Lacanian sense that I describe in Chapter 4 as constitutive of the "sexual child." In Piaget's preoperational stage (approximately 2 to 7 years old), children begin to evoke mentally things that do not actually occur. For example, they can imagine what will happen if a ball is dropped and are able to retrieve it if it disappears under the couch. Children can also represent the world in terms of symbols and can operate upon them at an intuitive level (long sticks can demarcate train tracks and roads). Spatially, they are still egocentric in that they have difficulty decentering themselves from any one aspect of a situation. None the less, children at this stage are able to operate within simple "topological geometries" and, as a consequence, may think of places in terms of "spatial primitives" such as proximity and connectedness. Piaget was at pains to distinguish his work from that of behaviorists who constructed children as separate from spatial stimuli. In his view, children actively participated in their environments but he none the less believed that space was a container for their activities. That space is an existent reality that contains the activities of the child is an assumption that continues to ground contemporary cognitive psychology (cf. Tversky *et al.* 1999).

At the concrete operational stage (approximately 7 to 11 years old), the intuitive constructions of the pre-operational period become stabilized into higher forms of mental representation. According to Piaget, children at this stage are now capable of linear thought. They can also abstract knowledge beyond self and no longer fuse or confuse their own point of view with those of others. Environmental relations, where place may be thought of in terms of linear projections of straight lines, are encompassed with notions of reversibility, composability and association, and are ultimately replaced by Euclidean metric relations (where the child can use a fixed coordinate system and distance measurements). At the formal operational stage, children are not only capable of linear thought but also discursive and logical reasoning. Reasoning is freed from "reality" in this stage so that children can abstract to new and novel contexts that they have not yet experienced.

Expanding horizons

Spatial coordination and expansion is a core element in Piaget's developmental ideas and, at the height of spatial quantification in the 1970s, geographers were quick to appropriate his model. Waves of empirical researchers weaned on Piaget gathered evidence to suggest that successive levels of cognitive development related specifically to the child's orientation in space, environmental competence and "mapping accuracy" as they became able to handle more and more sophisticated and abstracted geographic concepts (Moore 1976; Matthews 1984a, 1984b, 1986; Golledge *et al.* 1985, 1993, 1995; Downs and Liben 1987, 1991,

1997; Downs *et al.* 1988; Blaut 1991, 1997). Hart and Moore (1973), for example, discuss Piaget's work in terms of a qualitative change from ''action-in-space'' to ''perception-of-space'' to ''conceptions-about-space.'' One suggestion that relates to both children's spatial knowledge acquisition and their environmental mastery is the notion that competency increases as the child's horizons expand from the crib to the home, garden, neighborhood, city, nation and so forth. In his theory of environmental mastery, Erikson (1969, 1977) argued that competence comes through an expanding relationship between a child and three successive scales of environmental contact: the autospace, the microsphere and the macrosphere. The autosphere is an egocentric sensory environment or the ''first geography'' that then expands with age (Erikson 1977, 11). The notion of a child's expanding horizons articulates a particular kind of geography onto Piaget's developmental sequence. Robin Moore's (1986) ecological framework suggests nested fields or overlapping territories that also expands with age. He borrows heavily from biological metaphors to suggest that children occupy a particular kind of environmental niche.

Hugh Matthews and Melanie Limb (1999, 65) propose that scale provides a useful way of understanding children's environmental experiences. Arguing that children actively reach out from the self to a series of different spatial scales, they follow developmental work from Erikson and ecological work from Moore to suggest a nested hierarchy of experience from which a child's knowledge and competence grows. Matthews and Limb (1999) rightly point out that this perspective is but one model of child–place interaction and they are careful to note that it may not be an appropriate kind of model for children with different physical abilities. They also accept that children and childhood are not natural but social constructions and are quick to divest themselves of Moore's biologistic metaphor. That said, an enduring focus in this kind of research is to specifically isolate the processes of spatial knowledge acquisition and environmental competence while treating scale as a unproblematic linear extension from the micro-world of the self to the macro-world of the globe, and space as a sensible, verifiable and mappable reality.

Mapping the spatial child

A kid picks up *Winnie-the-Pooh* and makes complete sense of the map on the endpapers . . . without having the slightest instruction in map reading. Another opens *The Hobbit* and, although the map lacks a legend, is none the less able to follow Bilbo and the Dwarves across Wilderland. . . . What kid has a problem with the map in *Treasure Island*? With making sense of the map – and its two scales – in *Astérix le Gaulois*? Why is there so little . . . *resistance* . . . ? It is because the map is not apart from its culture but instead *a part of its culture*. It is because, as a map-immersed people, its history is our history. It is because we grow up into, effortlessly develop into, this culture . . . which is a culture of the map. . . . Since the culture is whole it doesn't really matter where we start (the clown stenciled on the crib is as good a place as any). We can plunge into it anywhere (which is where we enter it as children).

(Wood 1993, 143–4)

One of the more protracted debates in empirical scientific geography focuses on children's abilities to map. On one side of the debate, Roger Downs and his colleagues use Piagetian theory to argue for a gradual emergence of map under-standing in children that is not fully formed until the formal operational stage. They do not believe that mapping abilities are innate but, following Locke, they argue that the reasoning processes through which maps may be mastered are available to very young children (Liben and Downs 1997, 159). For Downs, the epistemic Piagetian child is an active, self-regulating agent, operating in and on the environment to *construct* knowledge and operational structures (Downs and Liben 1997, 179). On the other side of the debate, Jim Blaut and his colleagues argue that children entering school can deal with maps and have already apprehended macrospatial concepts (Blaut and Stea 1971). Blaut charges that the idea of Piagetian constructs encompassing children's development is at best mechanistic and at worst tautological:

Classic Piagetian theory tended to be very assertive about stages of development and rates of progress, in part because the typical form of argument was to start with a discrete intellectual "concept," then to find out at what stage or substage, and at what approximate age, it manifests itself in behavior.

(Blaut 1997, 172)

Blaut's basic premise is that all normal human beings of all ages in all cultures carry out mapping behavior, "it is therefore a natural ability, or habit, or faculty, 'natural' in a sense very close to the way language acquisition is 'natural'" (Blaut 1991, 55). Children of all ages, he asserts, engage in various kinds of mapping behavior and these abilities do not result from training or from incidental expo-sure to map-like stimuli, but rather from protomapping abilities which permit them to represent macro-environments. Using Noam Chomsky's (1965) theory of natural syntax, Blaut argues that if the human mind supports innate language acquisition, then it can also support innate map acquisition. These protomapping skills are manifest in toy play (where toys are assumed to represent reality reduced in scale), in movement (where mobility enables the child to ponder problems of the macro-environment and solve them), and in very young children's ability to read an aerial photograph. For Blaut, the physical context of a child's development, and their freedom to move within that environment and represent it in toy play and model-building, establishes the basis for natural mapping. Blaut (1991, 63) is not particularly concerned whether the source of mapping abilities is primarily an innate or primarily an early enculturation process in which experience and the social environment build upon extremely primitive inborn abilities. Rather, Blaut's point is that mapping behavior is universal skill which transcends culture.

Like Piaget, Blaut and Downs spent years developing tests to elaborate their respective theories about children's mapping abilities. One of the liveliest aspects of their discussion relates to Blaut's (1971) initial use of aerial photographs to suggest that even children as young as 3 years old can locate themselves and identify features from a "bird's eye" perspective. In later work, he and some colleagues test the same theory amongst 4-year-olds in a wide array of cultural settings (Blades *et al.* 1998). Downs insists that Blaut's empirical methods are flawed in large part because they cannot be replicated. In an attempt to do so, Downs *et al.* (1988, 690) conclude that young children have difficulty in maintaining scale and size relationships because they do not have the cognitive skills. Blaut (1997, 154) counters that the evidence from a small-scale photograph that was unfamiliar to the children such as that used by Downs and his colleagues is anecdotal at best. Many of Piaget's detractors, like Blaut, base their claims on Piagetian empiricism, arguing that his tests do not reveal developmental sequences but "measurement sequences" (Brainerd 1978; Donaldson 1978). Andrew Siegel (1981), in a search to find better questions to ask, concludes that distinct and sequential stages in a child's development of environmental knowledge may reflect, as much as anything else, stages in abilities to externalize spatial representations. But maps are spatial representations that are embedded parts of our culture. Children encounter maps and map-like forms everywhere, and from as soon as they are able to focus. Downs' and Blaut's conceptualization of maps is very specific, Euclidian, and related to only one type of behavior, spatial wayfinding, and one type of development, spatial knowledge acquisition. The questions that this raises, of course, derive from the extensive and rich geographic literature on the nature of maps and other spatial representations, not much of which appears in the Blaut/Downs debate.

One of my reasons for bringing up this thirty-year – and somewhat acrimonious – debate, then, is not to highlight its clearly irresolvable nature/nurture assumptions or to suggest that one set of empirical tests is better than another. Rather, I want to turn the debate around by thinking not so much about development and the nature of children, but about the nature of maps. For both protagonists, maps are relatively unproblematic representations of the spaces that contain children's activities. For Blaut (1991), they are culturally defined systems of symbols but there is a universal mapping impulse that leads to these mapping systems. They are models that represent in material form children's learning about place (Blaut 1999, 512). Similarly, Downs (1985) argues that maps are "realizations" of space in the sense that they make real something that was previously unattainable and in the sense that they help children (of a certain age) realize something that they had not understood by direct contact. For Downs and his colleagues, maps are "symbolic and spatial representations" whose "marks, shapes, and pixels are not referents themselves" (Liben and Downs 1997, 161). These are reasonable enough assertions but they do not go far enough.

The rather obvious point that I want to raise here is that maps and everything they contain are referents to systems of power-knowledge. By what they depict and what they omit – the presences and absences – maps represent particular worldviews and structures of knowledge. An important benefit of the rise of critical perspectives in geography of late is the deconstruction of these "techno-logies of power" (Gregory 1994, 7). The map is not distanced from culture and spatial wayfinding, and spatial knowledge acquisition cannot be understood as separate from powerful individuals and standardized cultural codes. The map is a powerful tool that hides as much as it elaborates, but it is also part of culture and, like everything else in culture (how to speak, how to use the toilet) its use is learnt. But when the map is read so that we may locate ourselves, we also locate – unconsciously and subtly – the map's history and politics. When a child finds her way home from school, for example, she probably reads the land-scape for signs of "stranger danger" as she has learnt to do from teachers and parents, as is taught by her culture. From *The House on Mango Street*, Sandra Cisneros' classic account of growing up in a poor multi-cultural American neighborhood, a resident child understands that

> those who don't know any better come into our neighborhood scared. They think we're dangerous. . . . They are stupid people who are lost and got here by mistake. But we aren't afraid. . . . All brown all around, we are safe. But watch us drive into a neighborhood of another color and our knees go shakity-shake.
>
> (Cisneros 1989, 28)

Returning to the scientific construction of maps and its relations to child development, what do we learn about the unproblematic use of mapping meta-phor to conceptualize children's acquisition of knowledge? It seems clear that cognitive scientists constitute children as developing Cartesian subjects rather than beings who live in specific places, in poor multi-cultural neighborhoods. This way of knowing subjects developed in conjunction with an Enlightenment individualism that continues to be inextricably tied to a specific concept of space and the technologies of power – such as the map, but also global positioning and geographic information systems – invented to deal with that space. Cartography is a science that standardized a certain form of representation during the Enlightenment, but it is also an expression of new forms of subjectivity that arose at this time and new technologies that allowed those new forms of subjectivity to coalesce (cf. Rose 1993; Kirby 1996).

Kathleen Kirby (1996, 46) asks "What kind of space, what kind of subject, does mapping (per)form?" As a way to document their knowledge of homes, local environments, neighborhoods or roots to and from school, children are often asked to sketch maps. This methodology is commonly used amongst geographers and cognitive behavioral psychologists (Matthews 1984a, 1984b, 1986; Spencer and Blades 1986; Golledge and Stimpson 1997). The idea behind it is that if children produce increasingly sophisticated sketch maps as they mature then there is evidence of the development of greater spatial cognition and mapping accuracy. Now, certainly, the argument for a child's cognitive map as an actual map – deriving from the work of O'Keefe and Nadel (1978) – fell into disfavor quite early on in preference for understanding cognitive maps as metaphors for the way the brain works to understand spatial information. That said, Golledge and Stimpson (1997, 238) argue quite convincingly that today's medical and computer emphasis on neural mapping is heralding renewed interest in the argument that the mind, like a cartographic map, has Euclidean spatial properties. Indeed, they contend that cognitive configurations constructed in two-dimensional Euclidean space correlate highly with Euclidean representations in objective reality although they concede that cognitive maps seem also to exist in multimetric and curvilinear spaces. The point to pick up from this is, of course, that no matter how many dimensions it has or whether it curves or droops, this space is always Cartesian. As such it is a space that is bounded and exclusionary, emphasizing absolute locations in abstraction over places and sites imbued by quirky and often unknowable contexts. As Brian Harley (1992, 246) notes, "context is stripped away and place is no longer important."

The philosopher René Descartes preceded Locke by about 100 years but it was upon notions of Reason derived from Cartesianism (and other more general philosophies of exclusion that categorized the early seventeeth century) that Locke's empiricism and analysis was founded. With a precision and clarity that escaped other philosophers of the time, Descartes outlined a project that was intended to lead thinking out of a miasma of anxiety and uncertainty to a "seductive Either/Or: *Either* there is some support for our being, a fixed foundation for our knowledge, *or* we can accept the forces of darkness that envelop us with madness, with intellectual and moral chaos'' (Bernstein, quoted in Gregory 1994, 72). Alternatively, Harley's critique of scientific cartography draws heavily from critical Foucauldian notions of institutionalized power and follows Derrida's methods of deconstruction. His critique eschews Cartesian logic and attempts to uncover what is hidden by standardization and precise measurement:

> The primary effect of the scientific rules was to create a ''standard'' . . . that enables cartographers to build a wall around their citadel of the true map. Its central bastions were measurement and standardization, and beyond there was a ''not cartography'' land where lurked an army of inaccurate, heretical, subjective, valuative and ideologically distorted images. Cartographers developed a ''sense of the other'' in relation to non-conforming maps.
>
> (Harley 1992, 235)

Kirby (1996, 46) transfers Harley's cartographic insight into the realm of the subject by pointing out similar emphases upon ''propriety'' and ''own-ness'' in the ''one-ness'' of the Enlightenment individual.[4] The history of standardized maps in Europe, argues Harley (1992, 241–2), was one of increasing detail and planimetric accuracy that ''gave a spiraling twist to the manifest destiny of European overseas conquest and colonization.'' This dovetailed with a further tendency in standardization to obliterate uniqueness in favor of stereotype. In part this reflected a European sense that the natural world should be neat and ordered. Now, it may seem a rather absurd leap to talk about colonization, maps and child development in the same paragraph. Nor do I wish to develop fully the interconnectedness of these ideas here, but it is curious to note that at about the same time that maps were standardized to enable colonization, containment and ownership, other shifts encompassed nature and children. As dikes and hedges enabled the enclosure of European fields, and the notion of the well-ordered English garden gained favor, Rousseau's and Locke's ideas of what constituted childhood began to foment into a coherent child-centered pedagogy.

Child-centered pedagogy and the ordered child

Child-centered pedagogy owes its origins to the work of Rousseau and Locke, in that they both advocated that children need nurturing and tutoring to realize their human potential. In the twentieth century, Piaget provided the structure through which this reasoning could develop. Rouseau provided the idealist assumptions, Locke the empiricist stimuli, and Piaget the theoretical structure upon which much of contemporary Anglo-American early educational practices are based.[5]

Roger Hart (1984) noted that in their attempts to construct the logic of intellectual development, Piaget and his students neglect the content of children's thought in terms of their affective development. With even more polemical bite, some sociologists describe Piaget's work as "an unholy alliance between the human sciences and human nature" where the child is seen as merely a "biological creature . . . with a grand potential to become not anything, but quite specifically something" (James *et al.* 1998, 17). In short, Piaget does not look at what children's interests are in the world; instead, his focus is on the construction of an ordered child. The countless experiments on the development of children's spatial thinking not only de-center emotional experiences as they relate to the social and physical environment but they also underestimate children's abilities. Piaget's structuralism posits a linear, hierarchical model and suggests a universal sequence of development, but as I noted in the previous section it is also entrenched in a mechanistic and disembodied philosophy of science that privileges reason and logic as the building blocks of knowledge. Despite these criticisms, Piaget's work became part of the foundation of a child-centered pedagogy that dominated the latter half of twentieth-century thinking on education and maturation. Because it concentrated on the organization of knowledge and understanding the developmental capability of the human mind in constructivist and interactionist terms, Piaget's work offered an optimistic and attractive alternative to deterministic organic and behavioral models that it rapidly eclipsed. But the use of Piaget's work for child-centered education suffered from premises that collapsed the infinite category of children into the more easily analyzed, educated and ordered "child." This monolithic conceptualization blurred the distinction between the individual and a socio-cultural context and thus essentialized and naturalized the growth process. If education was standardized to fit one conceptualization of the "child," then children were also constructed through surveillance and instruction to fit the educational curriculum.

Educational practices and control

Drawing on Michel Foucault's *Discipline and Punish* (1977), Valerie Walkerdine (1984, 155) locates the work of Piaget within child-centered pedagogy and educational practices. Importantly, these practices are manifest as much in quirky schoolroom geographies as they are in the materials taught. Walkerdine argues that child-centered pedagody constitutes not only a mode of observation and surveillance but also the production of children. Development in children is produced as an object of classification, of schooling, within a set of social and spatial practices. These practices are encompassed in what Shaun Fielding (2000, 231) calls a "hot bed" of moral geographies.

Walkerdine does not attribute the child-centered perspective to Piaget himself, but rather to the way his work was read, accepted and incorporated within an academy hungry to expand the application of rational thinking under the banner of humanism. In the 1960s and 1970s, Piaget's work was part of the legit-imization and production of practices aimed at liberating children. The task of educators was to graft context onto Piagetian theory. The formulation of the liberation, then, encompassed children as objects of a naturalized development that was completed with the understanding of rational scientific concepts. Walkerdine's main point is the failure of developmental psychology to meet the demands of public education's liberatory goals, and moreover the inability of practitioners to get beyond the monolithic concept of ''the child'' and attempt to deal with the practical contexts of children's lives and understand the ways that they really learn. Instead, ''many hours were spent demonstrating that Piaget was wrong here or there, that children were faster, slower, cleverer, or whatever than he had suggested, and that he had neglected this, that, or the other'' (Walkerdine 1988, 1). But his fundamental model of normative sequential development was never questioned.

Alternatively, Walkerdine argues that children are produced within discursive practices that cannot be accounted for in universal models of development or any kind of work that understands children outside specific historically and culturally located practices. Child-centered pedagogy, thus, has much to do with the control of space and its relations to the control of children.

Clearly, the control of childhood spaces functions through a combination of different institutions and devices and children's activities are regulated to a large degree. The child is able to emerge only through ''the disciplined, spatial implementation of the timetable which instills a regularity and a rhythm in all the activities and tasks of children'' (James et al. 1998, 55). Walkerdine shows how primary education environments, practices, and tools of analysis grow out of a pedagogy that is permeated by certain assumptions about child development. She examines the nature of child-centered pedagogy in formulating and evaluating educational practices, and how those relate to spatial environments:

> Particular disciplines, regimes of truth, bodies of knowledge, make pos-sible both *what can be said* and *what can be done:* both the object of science and the object of pedagogic practices. Pedagogic practices . . . are totally saturated with the notion of a normalized sequence of child develop-ment, so that those practices help produce children as the objects of their gaze. The apparatuses and mechanisms of schooling which do this range from the architecture of the school and the seating arrangements of the classroom to the curriculum materials and techniques of assessment.
>
> (Walkerdine 1984: 154–5)

Walkerdine (1984, 1988) is intent upon demonstrating that developmental psychology is premised on a set of claims to truth that are historically specific, and that are not the only or necessary way to understand children. She describes, for example, that the space of the classroom is also a moral space in which girls and boys learn radically different subject identities. Children are rewarded for individualism rather than individuality, for conformity rather than community, for compliance to authority rather than complaint (Wolfe and Rivlin 1987, 107). Schools are teaching and judging children by a set of criteria reflecting the status quo and designed to maintain it.

In a series of intensive studies of educational institutions over several years, Rivlin and Wolfe (1985) show that beyond academic goals (which usually translates to preparing children to pass exams), one of the main roles of the school is to teach children how to get along with others. This is not achieved, however, in terms of collective learning and socialization but rather by teaching the children to be "good citizens" and under the auspices of child-centered pedagogy, good citizens are children who conform to social norms and group behavior defined as appropriate by an adult authority. Children are taught how to accept this authority without challenge, rather than to define and implement their own group goals. Rivlin and Wolfe (1985) cite examples of how the classroom environment is designed to achieve this goal. In one alternative school, teachers separated desks under the assumption that any close proximity would result in antagonism and eventual disruption. Clearly, this kind of placement of desks is also used as a mechanism of control. Other, more subtle, forms of control exist in the classroom setting such as placing equipment or supplies where children cannot have access to them.

Teachers often gather children around them and concentrate activity in a small portion of the classroom. This control of behavior enables the teacher to monitor more closely the actions of the children. But in one example described by Rivlin and Wolfe (1985, 193), a teacher would trust some children and not others. The so-called high achievers and teacher's "favorites" were allowed to roam more freely and were not required to remain up front. They were allowed to leave the room more frequently and be out of her sight without raising suspicion. The children learn which of them is favored with less control and which have to comply to authority. Walkerdine (1988) shows how differential control of behavior and favoritism is often established along gender, class and racial lines. Girls, for example, are rewarded for docility and nurturing while boys are expected to enact the "naughtiness . . . validated and associated with masculinity" (1988, 229).

Fielding (2000, 236) points out that many of today's educational practices are self-referential. To the extent that "an overt moral 'framework' within which children and teachers are free to establish the best forms of teaching and learning in their particular space *appears* to generate . . . 'successful' teaching and learn-ing,'' that same framework ''throws up a whole suite of differing classroom and children's geographies'' that are problematic. Moral codes about where and how children ought to learn and behave raise concerns about how these are in actuality played out by children. If the major purpose of school is to socialize children with regard to their roles in life and their places in society, then perhaps it also serves the larger stratified society by inculcating compliant citizens and pro-ductive workers who will be prepared to assume roles considered appropriate to the pretension of their race, class and gender identities.

Social practices and resistance

Allison James and her colleagues point out that sociology's project has always been concerned with the developing child in that theories of social order and stability depend upon the development of predictable standards of action amongst partici-pating members. Potential participants, they aver, ''are always children and . . . the process of this inculcation is referred to as socialization'' (James *et al.* 1998, 23).[6] By focusing on universal notions of socialization rather than mutable and changeable social practices and ignoring differences such as gender, class and race, there is tacit support for relations of dominance because the production of these forms of identity is masked throughout the lifecycle. These are precisely the same forms of dominance that remain hidden in assumptions about the development of the spatial child.

Concerned with unmasking this form of deception, soviet psychologist, semiotician and pedagogical theorist Lev Semenovich Vygotsky and his followers formulated some of the first accounts of a socio-cultural psychology grounded in Marxism. Although there are connections between the liberatory premises of Vygotskian and Piagetian approaches (Downs and Liben 1993), the main distinc-tion between the two perspectives rests with the former's focus on socially constructed difference while the latter presupposes a universal form of spatial development. James Werstch (1985, 5) notes that the importance of Vygotsky's work was his decision to de-emphasize traditional Marxist concepts such as class consciousness and the psychology of fetishism and alienation in favor of a focus upon the nature of human activity, mediation and, in particular, the social origins of psychological processes that promote difference. His work provides an impor-tant underpinning for socio-cultural psychological development but, like Locke and Piaget, Vygotsky believed in universal human rationality and progress (Vygotsky 1987). The rationality that his work presupposes focused on ways children, and in some cases adults, learn to conform to social norms and, as such, work based on his ideas does not distance itself sufficiently from work based on other kinds of structuralism to be critical. A major dimension of criticism that remains absent from the work of Vygotsky and his followers finds focus in contemporary notions of social resistance.

The work of Paul Willis (1981) in his acclaimed *Learning to Labor* couches the lives of working-class youths in terms of covert resistance to schools, families and the larger structures of capitalism. Willis' study is of the schooling of British working-class males and their preparation for eventual waged labor. The impersonal structures that organize modern society must be understood, he argues, as being historically and culturally contextualized. With ethnographic methods, he explored the subtle nuances, forms of behavior, and manners of speech exhibited in the everyday lives of the twelve ''lads'' (linked by friendship and a mutual conspiracy to non-conformism) in a working-class school. By translating an abstract framework of Marxism into the everyday cultural terms of his working-class subjects, Willis is concerned with a tension between their knowledge of the system and the implications of their rebellion against it. The ''lads'' have a remarkable insight into the nature of capitalist exploitation but, in learning to resist the school environment, they establish the kinds of attitudes and practices that lock them into their class position. Willis also shows how this oppositional culture created from school experiences resonates with other critical locales. The home locale and parental perspectives show that the culture is generationally reproduced. The very way they talk and learn how to castigate the system ensures their place in a particular sector of low-level waged labor, foreclosing any possibility of upward mobility. The ''lads''' denigrating attitudes towards women (as well as West Indians and Asians) perpetuate a patriarchal control of reproduction as well as bigotry, racism and segmentation in the workforce. Willis argues that this outcome serves the needs of capital for an unskilled labor force and surplus population of unemployed while at the same time controlling the social reproduction of the system. But the results are not necessarily guaranteed because of the complex and conflicting relations of everyday life to an impersonal system that, under slightly different conditions, could subvert the reproduction of capital. Willis' study exemplifies the social reproduction of class positions amongst working-class adolescent males in the stratified social system of Britain in the 1970s. Some critics suggest that the issues he raises do not necessarily apply beyond Willis' ethnographic time and place, or more specifically, the degree to which they apply is merely an empirical question (cf. Marcus 1986). That said, Robert Everhart's (1983) exploration of working-class culture and resistance in an American school suggests that, at least in an attenuated form, Willis' themes may categorize a large part of growing up anywhere. Similarly, Linda McDowell's (2000, 2001) work with ''her working-class lads'' in two working-class British communities points to a continuation of Willis' speculations about the control of social reproduction. Using ethnographic work with male school-leavers, McDowell raises an important set of concerns about the way changes in social reproduction and the crisis of (working-class) masculinity is integrally tied to larger global transformations that propel labor market restructuring. From the practical accounts of McDowell, Willis, Walkerdine and others, it is clear that the concepts of childhood and adolescence as social constructions are reproduced and reinforced through sets of local beliefs and practices that are tied complexly to global transformations, and no amount of observation and theorizing will provide an unconstructed account of children's geographies.

The socially constructed child and post-structural accounts of development

Social constructivism has been around for a long time, but the notion of a socially constructed child as a valid focus of study is relatively new and derives from the work of Chris Jenks (1982), Rex Stainton-Rogers (1989), Allison James (1993), Cindi Katz (1986, 1991a, 1991b), and David Sibley (1995b) amongst others. Social constructivists argue that childhood as a category of experience is discursively produced. Modernist discourses naturalize childhood as a time of change that is directed by specific sequences of development. The essentialist part of these discourses attributes to development, behaviors and feelings that are actually derived from social differences. Anything that is not part of normative development is labeled deviant. From this viewpoint assumptions about larger social and cultural structures determining the outcome of development or the nature of childhood are suspended:

> If, for example, . . . child ''abuse'' was rife in earlier times and a fully anticipated feature of adult–child relations, then how are we to say it was bad, exploitative and harmful? Our standards of judgement are relative to our world-view and therefore we cannot make universal statements of value. What of infanticide in contemporary non-Western societies? Is it an immoral and criminal act or an economic necessity? Is it an extension of the Western belief in ''a woman's right to choose''?
>
> (James *et al.* 1998, 27)

There is no universal child or any form of generalizable process of development in this perspective. Child identity is always plural and there are a multiplicity of ways to know childhood. Although childhoods are variable they are also intentional, predicated upon social, political, historical, geographical and moral contexts.

James and her co-workers (1998, 147) argue that the social constructivist position is problematic because it tends to bracket and even dispense with the materiality of childhood. But this is not necessarily so. Social constructivists are intent upon uncovering the day-to-day experiences that embody children's lives and, of late, that concern has expanded to include issues of corporeality and materiality. It seems to me that theorists who hold a social constructivist perspective argue simply that references to childhood (including children's bodies, labor and sex) reinforce certain hegemonic norms. Of concern is that these norms suggest that children are naturally incapable of certain kinds of thought and action. It is only by passing certain developmental stages, for example, that children are able to read and understand maps or, for that matter, the codes necessary to find their way around a neighborhood. Such perspectives are not only hinged around Cartesian sensibilities but upon specific ways of connecting with the material world that either hides or universalizes morals, values, politics and culture. Social constructivism highlights ways of knowing, representations and material transformations as mutually constituting but there is also an important set of issues that relate to the culpability of those who study children's lives. James and her colleagues note that the social constructivist perspective demands a high level of reflexivity from its exponents which brings me back, in concluding this chapter, to considering the positioning of adults in the social construction of children.

The notion of a socially constructed child was generated in part from a crisis of representation in the social sciences from the mid-1970s onwards. This crisis was based on the realization that a quest for an authentic other is not achievable for any group but it is always unfulfilled by children. This crisis began with the work of some anthropologists (Marcus 1986; Clifford 1988) and quickly propelled itself through most of the social sciences including geography (Barnes and Duncan 1992; Griffin 1993). With its focus on texts and academic writing, the crisis revolved in large part around the misrepresentation of others. Put simply, there was concern that through most of the twentieth century, the field practices of social science and the writing of academics for the most part distanced the researcher from the researched to the extent that our writing said more about "us" than "them" and, read this way, the stories were not particularly attractive. Epistemologically, we confront the dilemma that cultural difference may not be translatable from them to us. Politically, we face the thorny problem of distinguishing between them and us when that establishes a powerful hierarchical dichotomy where we presume to speak for them. And, if we are persuaded by Freudian rhetoric (and I think we should be in this case), the tendency is usually to refer to them as "the other." This then raises the ethical question of the morality of speaking for others, but children clearly cannot always speak for themselves. John Bradshaw (1988, 211–14) argues that relationships such as those between young people and adults gain authenticity through detachment and effective separation, and we need to respect that distance. But the presumption of this distance is that what is constituted as childhood may well be our judgements as to what matters in being an adult. The other issue here is, of course, that most children are in contact with adults on a day-to-day basis and if we try to compartmentalize children's lives as separate from our own then, as David Harvey (1992, 303) puts it, we "dwell on the separateness and non-compatibility of language games, discourses, and experiential domains, and treat those diversities as biographically and sometimes even institutionally, socially and geographically determined."

There is, none the less, an evolving social theory of privacy and the private that does not negate Harvey's concerns about segmentation, separateness and geographical determinism (Morris 2000). Rather, it suggests a right to resist the normalizing powers of science and society, including maps. Privacy opens up a space of transgression that enables a special kind of reprieve from social control, a relief from power, and a removal of disciplining gazes. I will discuss this notion of privacy more fully at the end of the book, but it is worth pointing out here that it partially speaks to Bradshaw's concerns about respecting distance.

An alternative to viewing childhood as socially constructed is to view it as unstructured and undefinable, but such a perspective is paradoxical in one sense because everybody experiences childhood. Jordanova (1989, 6) argues that this accounts for "the profound ambivalence which informs our attitudes to children and which is relived when we become parents ourselves." And so, she goes on to note, to make children other to our adult selves we must split off a part of our pasts, a piece of ourselves. The question of how children are known, then, is about how we suspend a belief in taken-for-granted meanings (James *et al.* 1998, 27). Post-structural theorists point out that if we are intent upon "liberating childhood," we must embrace their resistances and transform our assumptions of what constitutes "natural" child development. But although this chapter is for the most part critical of natural predispositions to childhood and child development it is possible that by throwing out nature we may also be throwing out the metaphorical baby with some very fast flowing and mercurial bathwater.

And so I am not yet ready to give up nature. To turn the questions posed at the beginning of this chapter on their head, it might be appropriate to illustrate the ways society and culture work on and change biological bodies and how societal images of childhood are "embodied" in the corporeality of children, including their sexuality. This chapter focused on the ways relations between nature and childhood established and elaborated in scientific and popular discourses, but it also holds clues on the ways corporeality and sexuality are linked by their mutual exclusion from scientific discourse through the nineteenth century. What I am suggesting is that by focusing on development as a natural phenomenon and then simply mapping some social and cultural dimensions onto it sidesteps any kind of critical engagement with childhood as an embodied discourse which, of course, it clearly is. What is needed is a consideration of how the nature of childhood is embodied and sexualized. This chapter focused on the ways that science, nature, developmental theory and mapping are disembodied and distanced from corporeal and material considerations and points out that this is, at best, a partial story or, at worst, a deceit. The next chapter takes the embodiment of children as its main engagement.

Notes

1 Hospital data at the time revealed that diarrheal disorders constituted a leading cause – up to 25 percent – of infant deaths and it was assumed that mother's milk was the main cause (Gollaher 2000, 100).

2 There is some considerable controversy over the practical implications of Rousseau's work. On the one hand, Pestalozzi, through whom Rousseau's ideas were transformed into the public sphere of ordinary schools in Europe, "based his methods on the work of the good mother in the home" (Boyd 1962, 2). On the other hand, in the early nineteenth century Robert Owen, the Scottish philanthropist who established schools for the children of the workers in his New Lanark textile mills outside of Glasgow, developed forms of pedagogy that were left wing and progressive while at the same time pulling from Rousseau. Owen, for example, fought not only for universal education but also was an early advocate of women's suffrage.

3 The simulated cave from Chapter 1 was the most popular exhibit in the Children's Museum. The most popular part of the cave was a small room above the main room up into which young people could climb. From this perch they could survey the adults below. This observation may fit nicely with Appleton's theory, but perhaps children simply like to play at hiding and, presaging some arguments I'll make at the end of the book, perhaps the privacy afforded by the room establishes a space where children subvert the normalizing gaze of adults.

4 Kirby (1996, 46) also points out that subject's "imbrication in the developing social form of capitalism," a point I will develop further in later chapters.

5 The debt that Piaget owes to Rousseau is perhaps best exemplified by his 1918 novel about the semi-autobiographical searches for self of a young man, Sebastian, who bears a strong resemblance to Émile (Walkerdine 1988, 174).

6 Socialization is about how societies sustain themselves, but it is also about the transference of culture from one generation to the next. According to James *et al.* (1998), sociologists envisage the process in two ways. First, the *hard* way sees socialization as the internalization of social constraints and external regulation. The notion of a regulating system of social structure derives from the work of Talcott Parsons:

> What he also achieves is a universality in the practice and experience of childhood, because the content of socialization is secondary to the form of socialization in each and every case.
>
> (James *et al.* 1998, 25)

This structural perspective is countered by a second, *softer*, approach to socialization that is advocated by symbolic interactionists. James and her colleagues argue that although this perspective generated a wealth of ethnographic studies of small groups, its baseline was always adult interaction competence and so children were never thought of as anything more than ''becoming adults.'' Sociology's symbolic interactionism posits a structure to which children must aspire and, as a consequence, does not sufficiently distance itself from Talcott's structuralism.

3

LEARNING THROUGH THE BODY

The stories of childhood and child development from the previous chapter suggest that children's lives are commonly assumed to be rooted in nature. At one level, children's existence is conflated with images of beauty, purity, wildness and innocence and, at another level, a natural progression of development is assumed to accompany children's growth. With each scenario, it may be argued that the child's body is the primary site of childhood. The shape, arrangement and internal configuration of children's bodies are the focus of scientific study and a starting point from which modern discourses on childhood derive. These discourses suggest that, more than any other group, pressures are placed on children to conform to socially accepted norms of behavior. These pressures are derived in large part from the ways adults perceive the structure and physical maturation of children's bodies.

An outpouring of interest in bodies and corporeality in geography was witnessed in a surge of review essays in the 1990s (Longhurst 1995, 1997; Callard 1998; Nast and Pile 1998; Simonsen 2000). These invocations of the body are heavily shaped by the imaginaries of certain bodies – women, ethnic minorities, gay men and lesbians, and people with physical disabilities – rather than others. As Felicity Callard (1998, 399) remarks, the "overworked figure of the body is often presented as queer, hybrid or cyborg; it is rarely caught up with the abject, the abjected, laboring body." Young people's labor and their bodies are also largely missing from this discussion. I do not say this to belittle the struggle of women, disabled people and those with exceptional needs, and lesbian and gay individuals against normative expectations and the performance of stereotypical roles, but to raise questions of why young bodies are not present in these evolving discourses. Part of the reason is certainly related to the seeming lack of political will amongst children. Ironically, another part of the reason is to squash the emerging political will of adolescents. Other reasons relate to the implicit ties between bodies and sexuality, and the deep-seated and inviolable cultural values of Western society

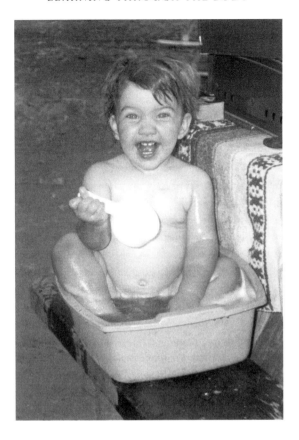

Figure 3.1 The baby and the bathwater. By throwing out an embodied sense of nature, we perhaps throw out the metaphorical baby with some very fast flowing and mercurial bathwater. My partner took this photograph of our daughter when we were sojourning with nature on a camping trip some years ago

Source: Peggy Beninger

that protect children from bodily harm and sexual improprieties. As suggested in the previous chapter, these moral issues although sometimes broached are rarely incorporated in scientific discourses that tend towards disembodied explanations. Understanding the representation of children's bodies, and its relations to their behavior and labor is of considerable interest for what I am trying to do here because it points to issues of difference and justice.

The pressures on children are different because they are expected to perform their childhood as well as (and often to the exclusion of) their sexuality and ethnicity, and their abilities are for the most part judged in accordance to physical size and stage of development. In addition, the pressures on children are different because it is difficult to tease out their reactions to others' expectations and the demands society places upon them. Unfortunately, these demands are often lofty when they are couched in a patriarchal discourse wherein children are expected to reproduce culture and society in the image of their forefathers.

By the early twentieth century, the emerging fields of cognitive development (as discussed in the last chapter) and psychoanalysis (the subject of the next chapter) joined with biomedicine in defining childhood as one of the most important projects of modernity. The point of this chapter, and why I use it as a fulcrum between discussions of child development and psychoanalysis, is to elaborate the hidden imaginings and moral enigmas that pivot around children's bodies. If considered at all in scientific discourse, children's bodies are hidden in a miasma of objectification and compartmentalization. The hidden imaginings turn, I think, on a recognition that the word ''body'' and the material object ''the body'' are all too often treated as obvious and unproblematic, requiring no explanation (cf. Longhurst 1997). Children's bodies, in particular, are simply growing into those of adults and the developmental literature alluded to in the previous chapter suggests that no matter where allegiances lie on intellectual growth and emotional maturation, physical growth is the mandate of nature not culture. The point of this chapter is to outline why this is not necessarily so and to illustrate an enduring story of child-like bodies that suggests deeper issues of desire.

The chapter begins with a discussion that highlights the recent geographic interest in bodies and corporeality. I then set out some of the ways science has come to know children's bodies. This discussion provides a springboard to a brief analysis of contemporary social science perspectives on children's embodied experiences with a particular focus on bodily changes and differences. I argue that a focus on young people's bodies requires also a call to destabilize the binary of sexual difference and the linear developmental sequences embodied through physical maturation. I broach the question of why adults impose such structures on children's bodies with the story of Mignon, the child-figure constructed by Goethe in the late eighteenth century, as selectively interpreted by Carolyn Steedman (1995). The indeterminate placing and fluid moral significance of this figure permeates European culture through the first quarter of the twentieth century.

I use Steedman's interpretation to selectively argue a compelling disciplining of children's bodies by adult desire and contempt. The discussion provides an important conduit to the next chapter that begins with the effacement of Mignon's sexual and developmental indeterminancy and the fixing of childhood through early psychoanalytic theory. I am particularly concerned about extending my previous discussion about the nature of childhood to a consideration of the embodiment of children and the seeming innocence of children as evidenced by what is problematically elaborated as a naïve corporeality.

Placing the body

A running theme throughout this chapter is that bodies are not just flesh and blood or an object that the mind uses. Ruth Butler (1999, 239) advocates "the body [as] an active and reactive entity which is not just part of us, but is who we are." Of late, bodies have become a part of the social agenda of Western societies to the extent that Kirsten Simonsen (2000, 7) argues that they are now recognized as a "cultural battlefield." Callard notes how much the body is used in contemporary Anglo-American fights for and over theory, suggesting that it:

> has been pursued, wanted, tugged, grasped, and torn apart. It has become a near endless resource with which one may produce compelling theory, harvest tropes, anchor one's desire to protect and articulate "difference," and, in short, find an absorbing locus for all manner of diverse, and often mutually antagonistic, projects.
>
> (Callard 1998, 387)

I argue that concern with children's bodies is not well placed in this evolving field of study but it is none the less critical because bodily growth, functioning and representation are issues that are continually recast in the social imaginary. The ability to address issues of children's bodies prefigures concerns about their sexuality and their labor but it also highlights the persistence of many other social problems and cultural issues.

Although geography has not been suffused with the wash of attention over the body that has characterized cultural studies or literary theory, an emerging literature of late suggests an important engagement (Bell and Valentine 1995; Longhurst 1995, 1997; Pile and Thrift 1995; Callard 1998). Simonsen (2000) sketches three arenas through which studies of the body have entered geography. First, the body as "the geography closest in" is an obvious endeavor. Second, and again obviously, is a need to understand the body as it is used to "other." Third, a focus on bodies is an attempt to transcend the Cartesian mind/body split.

The geography closest in

The problem with "the geography closest in" metaphor is that it suggests an expansion of self-awareness in a relatively limited linear progression away from the body as the self touches and comprehends larger and larger environments. As I mentioned in the last chapter, it further presumes a problematic metaphor of reaching out from the self to a series of different spatial scales. "The geography closest in" is a humanistic conceptualization of person–environment relations that belies the notion of space as a product of power relations that inhibit as well as enable. I argue elsewhere (Aitken 1999) that if it is possible to get beyond a naturalistic reification of scale then opportunities are offered to explore scale-related spatial narratives as stories about the creation of and resistance to boundaries and borders, including the ways bodies are bounded. Simonsen suggests that newer conceptualizations about "the geography closest in" might jettison any ideas of scale to focus on how bodies are made and used, the ways they labor, and how power is inscribed on and resisted by bodies. This perspective aligns more with Foucault's (1977) claim that socio-political structures construct particular kinds of bodies with specific needs and wants. From this perspective, bodies are primary objects of inscription for societal values and mores. A broadening of "the geography closest in" metaphor uses corporeality to encompass a material base for understanding the social construction of children as "other."

Other bodies

The construction of "other bodies" acknowledges not only the differences but also the power relations in embodiments. Bodies are inscribed by different race and ethnic backgrounds, by their age, sexuality and able-bodiedness, and by where they reside, work and travel. The body is central to how hegemonic discourses designate certain groups as "other" and how children are placed in each of these categories (e.g. female, obese, bespectacled, disabled, minority). Children's bodies are particularly susceptible because, as I noted at the end of the previous chapter, hegemonic norms suggest that they are naturally incapable of certain kinds of thought and action depending upon their developmental stage. To take just one example, most child-care manuals point out that toilet training is only possible after a certain period of maturation. It is usual in Western society to bind children's bodies in diapers or nappies prior to a sometimes intense and stressful period of toilet training so that they do not defecate where they please. But the practice of using a toilet – for both adults and children – is a cultural construction based on changing social practices and population pressures. Until quite recently, most adults defecated where they pleased. And so, in the contemporary West, at some time between 1 and 3 years of age, caregivers put young bodies through a stage where they are trained to sit on a large porcelain receptacle. A central topic of conversation amongst some parents, diaper-less children are displayed with great pride. Toilets could easily swallow a small bottom, and where does all that fecal matter go when it is whisked away with the swirling water? Clearly a potentially chilling thought for some young imaginations. Exerting further pressure on children and caregivers is a growing mandate between pre-school child-care centers in the United States requiring that their young charges be toilet trained. Some states require special licenses for care facilities that accept children who are not toilet trained, but many facilities simply do not wish to deal with diaper changes. For the working parents of these young children, then, the pressure to toilet train is more than social, it is an economic imperative.

Not only do age and growth give meanings to bodies but so also do sex, ethnicity, ''race'' and other differences, such as still needing a diaper at the age of 4. For older children, observable differences such as wearing retainers over teeth or glasses over eyes intersect to make meanings that are always imbued with power relations. Work on the body as ''other'' is clearly based on the notion that bodies are not generic but bear the markers of culturally constructed differences. By imprisoning the other in her or his body, privileged groups – particularly male, bourgeois, white heterosexuals – are able to take on positions as disembodied master subjects who make determinations about everything and everyone. And, importantly, they make these determinations from no place and no body.

Despite a critical exposé over the last decade or so, hidden positions of privilege remain difficult to breach. For the most part, those occupying hegemonic and hidden positions of privilege still set moral standards and conduct associated with the body, but what is particularly interesting here is how these standards work differently for children and adults. While the standards are positioned so that those who are judged cannot and/or will not meet them, they are set in such a way that they may become part of the aspirations of those who are young. Children are excluded from many moral judgements because they are embodied by discourses that foist a child-centered pedagogy on a socially constructed innocence or wildness. Through their bodies, children are seemingly exempted (or, rather, located differently) from the moral order until they can be marked as other or, with appropriate maturation, embraced. Understanding the embodiment of children requires that we sort out cultural and economic constructions from naturalized developmental outcomes.

The mind/body split

An understanding of ''the geographies closest in'' and ''the construction of other bodies'' clearly overlap and it is not necessarily useful to try to set them apart. Simonsen (2000) notes that both perspectives come together in a critique of the hegemonic dualisms that permeate Western culture. And so, third, a focus on bodies is also an attempt to transcend the Cartesian mind/body split and the ways it slips into other dualisms such as culture/nature, public/private, man/woman, constructivism/essentialism, adult/child and so forth. Of course, the mind/body dualism permeates a large swathe of Western philosophy since Plato, who believed that the mind dominated matter and that the acquisition of knowledge required that the body be disciplined by and subjected to the mind. But it was from a small room in a seventeenth-century provincial French town that René Descartes – in a lonely chamber, longing for release from the body's encumbrance (Zita 1998, 166) – developed a foundation for modern scientific knowledge that required the separation of mental discipline from the seemingly irrational and certainly unruly passions of the body. He inscribed on the body a modernist trope that structures corporeality as a mechanical substance reducible to itself. Descartes expanded knowledge, step by step, to admit the existence of God (as the first cause) and the reality of the physical world, which he held to be mechanistic and entirely divorced from the mind. This is almost complete dualism because it requires a separation of the material body from the inner self, and creates an ontological gulf that is bridgeable only by divine intervention.

Of course, this is only one way to fantasize about the body's matter. While the Cartesian project gives mind, masculinity, rationality and sameness priority over the body, feminity, irrationality and otherness, bodies are none the less always present although their role is complex. Simonsen (2000) notes that critical discussions in geography, and most of the other social sciences, focus on embodiments and discourses about the body with very little attention paid to the material body. Rather, priority is given to a deconstructive project that attempts to dismantle the Cartesian mind/body split or destabilize hegemonic notions of the body and dominant discourses on embodied identities. She notes that there is a certain amount of ambivalence towards material bodies and individuals' everyday interactions with their bodies. And, with the exception of Nast and Pile (1998), who note that representations of bodies cannot be understood outwith considerations of *place,* little geographic attention is given to how people interact with the world around them through their bodies.

In the balance of this chapter, a discussion of children's bodies as a scientific abstraction is followed by a review of some contemporary sociological research on children's experiences of their bodies. At the end of the chapter I reconsider the implications of Simonsen's concern about the research ambivalence towards the materiality of the body and suggest that what is missing is a focused appraisal of children's sex and work as embodied experiences. The call to understand the body is simultaneously a call for the fluidity of subjectivity and its relations to place and context, and so this chapter stands at odds with the last chapter, which focused on the ways child identity is fixed by some popular and scientific norms. Here I am concerned with the embodiment of childhood. In an attempt to not foreclose on the substance of the body, I set out to destabilize sexual difference and physical maturation, which is begun in this chapter with the story of Mignon and continued in the next chapter with an appraisal of psychoanalysis.

Studying children's bodies

The sum of technical writing about childhood increased from its slow beginning in the middle years of the nineteenth century to a virtual explosion of information on physiology that was foundationally and irreducibly focused on the body. A biological/anatomical body that is constructed as real, separable from mind and social construction, and reducible only into parts such as internal organs and skeletal frame provides the principal scientific model that became the foundation of modern paediatrics, and the scientific understanding of childhood. This foundationalist perspective predominates Anglo-American thinking throughout the nineteenth century and derives in large part from the Cartesian mind/body split. In Cartesian writing the disciplining of the mind is encumbered by the excess and uncontrollability of passions and desires linked to the body. By training young minds to pursue nothing but clear and distinct ideas, the Cartesian dream voids the chaos and mess of corporeality. The human body is seen as an aggregate of mechanical parts.

In late nineteenth-century science, functional morphology focused on physiological processes and practitioners within this field of study were keenly interested in disinterring the relationship between the structure and function of living forms. In short, the structure of a life-form (plant, woman or child) was the key to understanding how it functioned. An ideal plan or structure provided a template upon which all other structures were judged, and functional morphologists were intent upon a scientific search for this plan. Intensive study of the human body was presupposed by the notion that in its entire organization the body contained all the forces necessary for its functioning. At the methodological level, physiologists believed that all the activities of the human body could be explained in terms of its material composition and the interaction of its parts.

At this time, physiological thought (from which biology and studies of physical development derive) was influenced heavily by evolutionary theory. Social physics exercised a strong influence on physiological growth studies and ultimately eugenics (Quetelet 1835, cited in Steedman 1995). And so, aspects of physiological theory presupposed the notion of an ideal man and a superior race conceived in modern (post-Darwinist) terms. The importance of children in all of this is that, in terms of bodily growth, functional morphologists believed that an ideal structure resided within the organism and it was this structure that enabled transformation and metamorphosis. Children's bodies were appropriate sites of study because they were closer to the ideal form.

In many ways, these ideas are part of the foundation of nativist and empirical understandings of child development discussed in the previous chapter but here I want to focus specifically on notions of children's bodies as ideal forms. Growth for the functional morphologists comes from an interior ideal type that expresses itself through exterior developments. It was important for scientific study of this ideal that children's bodies be allowed to develop "naturally." At the same time, young people were contained and structured through the widespread nineteenth-century practices of coddling infants (literally binding their bodies) and using children and youths in mines, chimneys and textile plants to the extent that their bodies were deformed and misarranged. More on this later.

The notion of an irreducible interior life-force that developed naturally rather than simply grew from a scaled-down version is found in Rousseau's thought that the intellectual development of children retraced the steps of civilization, which developed through the work of Lamarck into recapitulationism. The intellectual nexus of these ideas was discussed in the last chapter. The important point here is that biologists understood from the early part of the nineteenth century that organisms do not simply grow and that adults are not simply children who get bigger. Rather, children develop and transform. But for functional morphologists, developmental sequences take place in a geometric transformation of a basic plan or design. Previous thinking focused simply on the plan getting bigger. Foucault (1970, 228) sees this change in thinking in the late eighteenth century as an archaeological shift in understanding from one where living things are categorized and conceptualized to one where the notion of "life" and bodily changes become indispensable to the ordering of natural things: "It was essential to be able to apprehend in the depths of the body the relations that link superficial organs to those whose existence and hidden forms perform the essential functions." Carolyn Steedman (1995) notes how this sea-level change focused efforts on the physiological development of young bodies through the nineteenth century. Her reading of popular physiological and educational texts of this time suggests that parents and doctors were directed to scrutinize outward manifestations (e.g. heads, tongues, gestures, cries) that may guide watchful observers to another level of observation – circulatory, respiratory, digestive – that was wholly interior. It may be argued that a sense of "interiority" – with an accompanying foundational and embodied spirit – emerged at this time and evolved slowly to contemporary psychoanalytic and therapeutic pre-occupation with adults searching for their "inner child."[1]

It is interesting to trace, then, emerging medical views of children's bodies and the embodiment of children in the adult psyche because they speak to larger child–adult relations. The emergence of modern biomedicine entailed the creation of an anatomy of pathology with a particular focus on disease and pathologies within specific sites of the body. At first there was little distinction made between the bodies of adults and children. Only gradually were distinctions teased out between adult and child versions of particular pathologies. When distinctions were made, Steedman points out that metaphors used to embrace understanding of the infants' interior body were horticultural in form and the methods of observation were derived from Locke's empiricism, but they hinted also at what later became child-centered pedagogy: "What success should we expect of a gardener, who was totally ignorant of the nature of plants and the relation that subsisted between them and the air around them or the soil in which they grew?'' (Smiles 1838, quoted in Steedman 1995, 71). The important point that Steedman notes is that childhood was gradually isolated from adulthood through nineteenth-century biomedicine in its depiction of the nature of the pathological interactions taking place within the child's body. Children's bodies functioned in fundamentally different ways from those of adults and that functioning was distinguished on the basis of biomedicine's understanding of how children grew. Attention was focused in particular on the vulnerability of the growing period and the need to tend young children like young plants. Through the nineteenth and twentieth centuries, concern with the diseases of children evolved into a general concern with children's health and development, but it was not until the early twentieth century that the distinct medical focus on pediatrics emerged.

In his study of British medicine, David Armstrong (1983) argues that the emergence of pediatrics is a direct result of societal changes in the relationship between adults and children although there was clearly biomedical complicity in these changes. Pediatrics are possible only after a medical shift away from trying to understand the diseases within children to a more coherent focus on defining diseases that are peculiar to children. Physiological functionalism developed through the nineteenth century by focusing on the internal working of bodies that precluded distinctions by age or stage of development but by the beginning of the twentieth century there was focus on the peculiarity of children's bodies as distinct from those of adults. Armstrong argues that childhood bodies were not so much discovered as invented in the context of British medicine as pediatricians increasingly claimed that their object of attention was ''normal'' childhood as opposed to childhood pathologies and diseases. It was assumed that physiological limitations of its members determine the state and development of society and social relations, and so children's bodies were to be cared for, regulated and constrained out of fears that they, society's future, might not be fit to take their adult place in society. Armstrong argues further that pediatrics pioneered the shift in medicine from a concern with isolating disease in the body to a concern with the body in its social context. Using Foucauldian geneological techniques, he demonstrates a sharp shift in early twentieth-century British medical thought towards the surveillance and monitoring of children. The motivation for observation was to ensure the health of future populations.

One of Armstrong's arguments is that the discovery of infant mortality as an indicator of a population's health suggests a new way of thinking about social reproduction and creating a healthy and productive adult workforce. Armstrong argues against infant mortality as a historical constant. Rather, he sees it as an invention of the end of the nineteenth century, the product of a certain way of thinking about the deaths of children as part of larger notions of development. Armstrong points out that the data for calculating infant mortality existed for decades prior to the creation of an index in 1877 that basically established the health of a population by counting the number of children who died in their first year. David Gollaher (2000, 100) notes that the North American context was similar, with the American Medical Association instituting a new section of pediatric medicine in 1880 after public health agencies began highlighting shocking rates of infant mortality in urban areas.

Today, infant mortality is often regarded as the single best indicator of a nation's "quality of life." Most demographers agree that, in general, the lower the infant mortality rate, then the better the health and social environment of a society.

Although the United States spends considerably more of its GNP on health care than other industrialized nations, its infant mortality rate consistently exceeds other "developed" nations. At the beginning of the century the United States had an infant mortality rate of 100 deaths per 1,000 births, a rate about equal to the rate in many contemporary countries in the global south. This rate declined fairly consistently until a mid-1970s value of about 15 deaths per 1,000 births. Since then, the US rate of infant mortality has improved only slightly whereas the rates of other nations in the global north fell markedly in the last quarter of the twentieth century. That this is of great concern in the United States highlights a larger debate on the intrusion of health surveillance into the private sphere:

> Housing, nutrition, hygiene and poverty became the analytic lines through which the domestic was brought from the private into the public domain. The relationship between infant and mother . . . rapidly became entangled in the web of analyses of domestic life . . . the infant mortality rate itself became, in a reflexive moment, an important indicator of social well-being.
>
> (Armstrong 1986, 213–14)

From this intrusion into the private sphere, a dangerously racist (and geographic) imperative emerged. The official rhetoric goes something like this. According to the United States Department of Health and Human Services (1986, see also Sanders and Mattson 1998), part of the explanation of high national infant mortality rates resides with the number of low birth weight babies. A weight of under 2,500 grams significantly lessens a child's chances of survival in the first few months, and is directly related to the quality of prenatal care. American minority babies generally fair worse than white babies in this regard. African-American babies, for example, are twice as likely to be of low birth weight as are white babies. According to official statistics, even when the variables maternal age, marital status and education are controlled, African-American babies remain twice as likely to be of low birth weight or to die. Government researchers are perplexed by the reasons for these disparities, but speculate on the effects of a constellation of "hard to measure" factors such as the childhood experiences of African-American women that unduly influence child-bearing. In addition, rates of infant mortality are generally higher in the central core of large cities where many minorities live. Residence in a rural area, and particularly in a band stretching from South Carolina through Arkansas, may also heighten risk for infant mortality (Hale 1990). Clearly, there is a geography to these ephemeral and dangerously essentialist speculations.

The importance of infant mortality rates is embodied by larger geographic and national concerns but there is another related example of relations between the rise of pediatrics and the control of children's bodies that is worth mentioning. The practice of infant male circumcision for health rather than religious reasons emerged in the United Kingdom and the United States in the second half of the nineteenth century and is directly related to the increasing influence of pediatrics. In the United States, hospital data from the late nineteenth century revealed that 25 percent of infant deaths was due to diarrheal disorders. Searching for ways to prevent gastrointestinal illness, many pediatric specialists recommended removing the foreskins of newborn boys. Gollaher (2000, 100) cites a widely read article of the time by J.A. Hofheimer that reports success in using circumcision to cure both fecal incontinence and constipation. It was from rhetoric of this kind that circumcision gained wide acceptance and became a familiar medical practice in Europe and North America. By the middle of the twentieth century it was regarded as a benign surgery that helped prevent urinary tract infections and diseases. Circumcision was certainly believed to have little consequence to issues of manhood or morality. It is possible to argue, however, that what was hidden by the familiarity of the practice was the maintenance of its roots as a basis for moral purification.[2] Control of bodies is about control of social practices and it is no mere coincidence that the widespread acceptance of the circumcision of boys dovetailed quite nicely with the late-Victorian concerns about masturbation. It is not difficult to argue that what was being controlled by circumcision was not male health but male sexuality. Sexual purity (i.e. abstinence from masturbation or intercourse outside of marriage) is part of the larger construction of good health. Constructed as subversive, aggressive and dangerous, men's sexual impulses could not be contained by the new forms of labor produced by industrialization or the supposed domestic security of the new private family. As Gollaher (2000, 101) avers, "just how imperfectly the bonds of work, culture, and society held male lust in check was apparent in cities, with their rising rates of illegitimate births and epidemics of venereal disease." The pain of circumcision, for example, was thought by many to eradicate masturbation and other secret vices which led to other improper sexual urges. In the United States, John

Harvey Kellog – the surgeon who gained fame for his obsession with dietary fiber and whose name is associated with a wide range of popular breakfast cereals – recommended performing circumcision on boys "without administering anesthetic, as the pain attending the operation will have a salutary effect upon the mind, especially if connected with the idea of punishment" (Kellog 1888, cited in Gollaher 2000, 103).

Gollaher (2000, 104) points out that there were no medical data at the time to suggest causal relations between circumcision and improved health or nutrition.[3] Notions of control and punishment, then, are suggested by the widespread Western practice of cutting boys' genitals.[4] I will return to this issue in the next chapter when I raise the specter of sexual control, but before I close the discussion of male circumcision here I want to refer to an interesting speculation made by Gollaher in the preface to his history of circumcision. I have not discussed female circumcision here, but its history is equally as long as its male counterpart. For reasons that are not entirely known but perhaps relate to the belief that women's sexuality can be controlled by other means, it is not practiced much outside of the global south.[5] Gollaher (2000, xiv) argues that if male circumcision were, like female circumcision, confined to so-called developing nations, then "it would by now have emerged as an international *cause célèbre* stirring passionate opposition from feminists, physicians, politicians, and the global human rights community . . . [male circumcision] is so deeply embedded in certain cultures and worldviews that it is hard to recognize it for what it is." An interesting geography of control arises that clearly ties bodies to places and cultural practices.

A rhetoric of normalcy and what constitutes a healthy population pervades the pediatric literature. By the beginning of the last century, this is usually expressed as a concern for creating positive health effects rather than the nineteenth century's focus on the treatment of disease. Iris Young highlights the notion of the scaling of bodies as they are contextualized by larger national concerns:

> In the nineteenth century in Europe and the United States the normalizing gaze of science endowed the aesthetic scaling of bodies with the authoritativeness of objective truth. All bodies can be located on a single scale whose apex is the strong and beautiful youth and whose nadir is the degenerate. The scale measured at least three crucial attributes: physical health, moral soundness, and mental balance. The degenerate is physically weak, frail and diseased.
>
> (Young 1990b, 128)

The important nineteenth-century shift for what I am arguing here is that children's bodies are marked empirically as a something that is manifestly present, but they are also manifest as the dominant signifier and a locus of control for adult society. And, disturbingly, government rhetoric around issues of normalcy suggests an added dimension that points to greater control. What shape does this control take and how does it fashion children and issues of childhood identity? Missing from most contemporary theorizing about childhood identity is a coherent understanding of how children's bodies, which are assumed to grow naturally, are in fact shaped and structured and how children come to understand their own bodies.

Feeling normal

For the most part, and until very recently, social science studies of how young people come to know their own bodies rested fitfully on foundationalist perspectives. That is, the child's body at its base is assumed to exist unproblematically as a reality that is known solely through biological and medical science. The job of exploring the lives of young people through child-centered pedagogy, for example, is to bring their evolving bodily awareness up to a certain standard. Allison James (1993) points out that until recently, and sometimes continuing today, psychometric testing revealed what stage of development a child had reached by whether or not they could label or draw the location of different body organs. The conclusions of these studies parallel closely the ways Piagetian theorists underestimate children's spatial and other abilities: ''Inevitably, children's knowledge appears as an incorrect or faulty version of the supposedly 'correct' (adult) version, although some investigators declare themselves 'surprised' by how much children seem to know'' (James *et al.* 1998, 149). From this foundationalist perspective, adult knowledge is fixed and is the standard to which children must aspire. Children's views of their bodies are linked to developmental stages that are universal. Ruled out from these generalities are notions that children's bodily awareness is contextual, fluid and contingent upon social relations and cultural embeddings.

The cultural stereotypes of what constitutes a normally developing child body assume great importance for both children and their caregivers. Brandtstädter (1990) suggests that normative developmental progress occurs precisely because widely held beliefs about children's bodily changes serve as the grounding rationale for the actions of parents. Gergen and his colleagues (1990) demonstrate empirically that close relationships exist between common, and sometimes misconceived, beliefs about developing bodies on the one hand and the mother's role in child-rearing on the other. German and American women in their study implicitly believed in an orderly sequence of growth and development. Given the similarities in views between mothers and non-mothers and between men and women, Gergen *et al.* (1990) conclude that these conceptions are cultural belief systems rather than any identifiable "true nature of the child." Moreover, not only did these belief systems affect the mother's child-rearing practices, but they also related to other conceptions that the women had of family structure and home/work relations. A potential problem of this work is an over-emphasis on the social and cultural domain of learning and identity formation without appropriate consideration of the bodily experiences of children themselves.

James and her colleagues (1998) overview studies from sociology that use ethnographies and participant observation techniques to provide an understanding of the contextual embeddedness of children's bodily awareness. For example, studying children with cancer led Bluebond-Langner (1978; Bluebond-Langner et al. 1991) to note that by reading the decline of their own bodies and those of others, as well as the body language and worried faces of nurses and caregivers, children accumulate knowledge about the progression of their disease. Changes in their bodies (e.g. hair loss associated with chemotherapy) provided important ties with other children in the same context that enabled the development of communities of shared identity and mutual support. When surrounded by healthy children, the children with cancer often isolate themselves because of teasing and stigmatization.

From an ethnographic study of children's bodily awareness, James (1993) notes five aspects of the body that focus on normal development: height, shape, appearance, gender and performance. Deviations from normative notions of how these aspects are constituted tended to produce extreme anxiety in her participants. Importantly, children's embracing of these features for their own sense of identity is paralleled by how they use them to identify children as other or deviant. Being "fat" or "skinny" were distinguished as divergences from the norm and provided bipolar identities rather than mediated positions along a continuum. If a child was fat or skinny they were deviant and both classifications may be treated as "other." James points out, however, that children's identification with their bodies is not necessarily fixed. For example, in the later stages of elementary school some children may think of themselves as "big" in relation to those in kindergarten but when they move to Junior High School these children become "small" once again. This fluidity in children's understanding of the relationship between size and status, James identifies with a certain "edginess" about body meaning. Although her examples may seem trivial, the important point that she makes is that childhood is a crucial resource for establishing identity precisely because of its unstable materiality. Questions raised by work such as this revolve around how much children's bodies are fixed by adultist notions of what it should be. The shaping and structuring of children's bodies takes an interesting turn at adolescence, when they are perhaps the most unstable materially and when changes are perhaps the most evident.

Puberty and the staging of bodily changes

Adolescence is constructed around body changes. It is a "stage" designed to encourage and, by so doing, survey and control, rebellious behavior. James Kincaid (1992, 69) argues that "childhood" is under somewhat better control because, as an institution, it is about learning rules whereas adolescence is about nominal revolution, and coming to terms with bodily changes and uncontrollable erotic feelings. The teenage years are an explosively emotional time for boys and girls when bodily changes are sometimes difficult to keep private. As Barrie Thorne (1994, 142) points out with her work in Californian working-class schools, personal bodily changes and sexuality for teenagers are the stuff of public commentary:

> Some of this commentary, like bra-snapping, takes ritual form. Once in a classroom and several times on the playground I saw a girl or boy reach over and pull on the elastic back of a bra, letting it go with a loud snap followed by laughter. . . . Kids are often curious about one another's bodily changes, which they may transform into public news.

The hidden, or at least ambivalent, sexual meanings of childhood become overtly sexual with the bodily changes of teenage years. The repression of sexuality prior to pubescence, and, indeed, the word "puberty," is an invention of modern Anglo-European culture.[6]

Adolescence was conceived as a "distinctly sexual" stage with its construction by Hall (1904) and since that time the near universal categories of adolescence and youth are characterized mostly by hormonal and biopsychological changes which occur during puberty (Hyams 2000, 637). Kincaid (1992, 70) makes the interesting point that the dividing line between childhood and adulthood found an almost universal form of delimitation, at least by the turn of last century, with bodily changes and a specific "fluid" rendering of the interior "leaking out." The commencement of menstruation became the borderline between girlhood and womanhood and the tendency of nocturnal emissions, perhaps a little less clearly, signified manhood. Kincaid goes on to argue that today most writers recognize adolescence as a social construction and not a natural state, pointing out that it is an institution that has proved most troubling since its manufacture.

Teenage pregnancy constitutes a seemingly recent troubling aspect of adolescence. Although the notion of women controlling their own bodies is central to the pro-choice/pro-life debate, both sides of the debate agree that teenage pregnancies are problematic. But why? Given the historical tendency for women to conceive in adolescence, the stigma attached to teenage pregnancies is a very recent phenomenon with an interesting local geography. In a study of young Latina women in Los Angeles, Melissa Hyams (2000) argues that girls' bodily changes are embedded in the larger society's anxiety-filled knowledge about young women but they are also constructed locally. This knowledge presupposes that desire is reserved for men and boys and, if imagined for young women at all, it is done so only in terms of complete loss of control. The discourse of desire, preparedness and responsibility around bodily comportment is missing in the social construction of young women. Hyams argues that in the socio-spatial context of Los Angeles, the discourse of anxiety-filled knowledge focuses on the young, poor, Latina woman and a body that is explicitly sexualized and implicitly racialized. For the young women that Hyams interviewed, learning about responsibility for modesty and chastity is central to the image that constitutes their high-school experiences. For her participants, sexual intimacy in a "serious" relationship did not threaten a young woman's reputation, but getting pregnant suggests "immaturity," deceit (by the boy) and/or loss of control. This may not threaten her reputation, but all Hyams' participants agreed that it damaged her future. In addition to this kind of action or inaction, these young women learned how to manage their images by bodily comportment and controlled displays of affection. Covering up, and being "calm" rather than "goofy" around boys suggested a maturity that would not elicit sexual desire. Hyams' participants embody a space of ambivalence between the fear of drawing attention to themselves through appearances and the desire for that attention. Where and how to get attention without conveying a wrong message of sexual unboundedness proscribed a series of subtle resistances. For example, long zipper-fronted sweat-shirts enabled some girls to expose when appropriate and otherwise cover up midriffs, cleavages and thighs. The zippers enabled some autonomy over self-representation and by wearing them, the young women push against certain positioning boundaries. That said, Hyams concludes that although some are able to gain a sense of themselves as competent, most of her participants' engagement with normalized and sexualized notions of adolescence left them feeling vulnerable and passive. Similarly, Hyams points out that many of her participants had significant anxiety over their wish to conform to codified sexual norms and voiced these anxieties in terms that related to failure in school.

Hyams' participants derided "hoochie mamas" as young women who were "popping out everywhere" and who perhaps flaunted the school dress code. Importantly, Hyams (2000, 648) argues that sometimes what was constructed as flaunting was in actuality predicated simply by young women who have "too much" to show rather than who "show too much of what they have." These girls represent a sexualized female form that embodies danger and desire. One relatively voluptuous girl who considered the way she dressed as "feminine" was characterized by other girls as a hoochie mama and by one of her teachers as someone who doesn't participate in class and "would end up pregnant."

Marked by her choice of clothes that suggested fuzzy sexual boundaries, this girl was also marked by inevitable teen pregnancy and academic failure. If "hoochie mamas" and their attendant images of victimization, loss of control and academic failure represent the lower limit of the schoolyard moral economy, then the upper limits – signifying modesty and chastity – are represented by Mexican women. Although most of Hyams' participants had never crossed the border, stories from their mothers, grandmothers and aunts suggest an image of purity, sexual propriety and respect from boys. For these young women in Los Angeles, their counterparts south of the border were an idealized, and unreachable, moral apex through which their sexual identities were racialized and schooled.

The important point about Hyams' work with Latina women for what I am arguing here is that adolescence and schooling coincides with the bodily changes that occur around the inception of puberty. But it also coincides with the expectation of parenthood and a consequent passage into the adult realm. At a larger societal level, adolescence as a "sexual stage" between childhood and adulthood was meaningless until lifespans lengthened, formal education was required beyond the pre-teen years, and pregnancies were delayed. Constructed as an "age-stage" prior to adulthood, adolescence is a transition to "social, economic, and psychological independence, monogamous heterosexuality within marriage, parenthood, and full-time employment. Failure or the threat of failure to achieve 'normal' adult status is pathologised as deviant, delinquent, or deficient" (Hyams 2000, 638). Hyams' work suggests that an important local geography is missed if the focus is solely on developmental stages, body changes and larger societal changes. There are important implications from studying local places that I will return to later, but it is not too early to point out that this tactic for knowing young people is predicated upon the grounded and material conditions of lived experience. One remaining question that must be raised before I can get to that discussion, and its raising takes the balance of this chapter and the next, is why do adults need to shape children's bodies?

Mignontage: the shaping of bodies

Human life is . . . but like a forward child, that must be played with and humored a little to keep it quiet till it falls asleep.

(Temple 1992, 413)

The classical age discovered the body as an object and target of power . . . the body that is manipulated, shaped, trained, which obeys, responds, becomes skillful.

(Foucault 1977, 136)

I want to argue that the forward child, habitually disposed to disobedience, adversity and opposition, is an appropriate metaphor for the resistance that many post-structural commentators point to as a necessary condition for subverting Cartesian logic. My enduring point in what follows is that seemingly benign forms of placating willful children are also forms of imagining their sexuality and growth. The metaphor that seems most helpful to understanding these largely hidden imaginings is the child-figure Mignon. Although the story of Mignon sank into relative obscurity in the latter half of the twentieth century, I use Carolyn Steedman's (1995) elaboration of its main themes to unpack some of the ways adults imagine children's bodies.

Phillipe Ariès (1962) attributes the French word *mignontage* to the practice of coddling infants. To "coddle" is to treat a child with excessive indulgence or pampering. Today it is morally acceptable in many circles to pamper children to the extent that their appetites, tastes and desires are catered for at least to some degree. It suggests a kindly or excessive lenience in yielding to wishes, inclinations, or impulses, especially those that are directed by the seeming chaotic passions of the body. Ariès saw mignontage in early seventeenth-century Europe as a feminized attribute of childhood wherein women nurses (and some men) coddled and pampered the children in their charge not only to placate them but also to indulge their caprice as entertainment. The English derivation of a "minion" as an obsequious follower or dependant comes from the Old French word *mignot* or *mignon*. This seeming feminization of childhood arose just prior to some (exclusively male and reactionary) pedagogic attention to the development of a child's reason from the late seventeenth century onwards. This latter focus on child-centered pedagogy was focused on the Cartesian mind/body split and inspired by Lockian empiricism discussed previously, but what I want to focus on here is the importance of *mignontage* to a quirky and partially hidden understanding of childhood. This imaginary is traceable from Goethe's work in the eighteenth century through to the early part of the twentieth century. The imaginary is important because, like the scientific study of children's bodies, it is indelibly linked to ways that adults control children's bodies and how that control reflects adultist needs. The widespread practice of coddling amongst Puritans in sixteenth- and seventeenth-century England was also about placating and pacifying infants, but it was far removed from yielding to a child's wishes, inclinations, or impulses. Rather, it required the tight binding and constraining of children's bodies, yielding a material control over their supposed propensity to willfulness.

Mignon is encountered as a character in Johann Wolfgang von Goethe's *Wilhelm Meister's Years of Apprenticeship* (1796). The book became the prototype of the German novel through innovative character developments and, as such, provides an interesting representation of the child-figure. Although the term *mignon* (read as minion) appears before this time to characterize Henry III of France's pederastic court entourage, most scholars agree that it is Goethe's personification of the term as a child-figure that captures popular imagination through to the early part of last century. Goethe studied law at Leipzig (1765) and at Strasbourg (1770–1), where he began his lifelong interest in plants and animals. One of his lasting influences from this period was Rousseau who appealed to his mystic feelings for nature. What is interesting about the development of the child-figure in *Wilhelm Meister* is that although Goethe attends to her mystic sensibilities, Mignon is none the less also a product of adult passion and desire.

The meaning of "mignon" carried at Goethe's time was that of fondling, little one, sweetheart, youngest one or little favorite (Steedman 1995, 27). In the gendered origin of the word, it is a masculine adjective that also has a female form. Goethe personifies the word in the form of a stunningly beautiful androgynous child acrobat abducted by a troupe of rope-dancers and deformed by their violent beatings so that she exudes a physically dislocated but none the less erotic aesthetic when performing. Androgynous and a cross-dresser (Steedman points out that Henry III's minions were derided by members of his court for wearing effeminate clothing that sometimes parodied their master's attire), by the time of her death it is clear that she is female. Her Italian origins helped establish Mignon's exotic "otherness." This figure is not grotesque in the same sense as Victorian England's John Merrick (a.k.a. the Elephant Man, Montagu 1971), but she is none the less haunting in her awkward movements, weird postures and limb movements. The child-figure represented by Mignon evokes great pity but there is also erotic investment in her dancing and demeanor. Steedman (1995, 28) points out that adult male reaction to her or him is truly a voice of desire and in this sense, Goethe's representation pulls heavily from Henry III's pederasty.

That Mignon's dislocated limbs are the result of adult violence is important because it removes from this child-figure John Merrick's "natural" disfiguration. It adds an element of culpability and guilt over her lived experience and ultimate death. A contemporary critic of Goethe's work thought it appropriate that Mignon's death be attributable to "the world's failure of understanding," highlighting "practical enormity and terrible pathos" (Steedman 1995, 26). Mignon is also variously represented as autistic and underdeveloped, accomplished only in her dancing, musical and singing abilities – a masque and spectacle for the enjoyment of an adult audience. In this sense, Mignon is enigmatic, self-enclosed and detached from the adult world and its ways of knowing. A form of this detachment is Mignon's preoccupation with her inability to read maps or to fully understand written language. In Goethe's original representation Mignon desperately wants to understand how maps work, what form of representation they constitute, but she cannot perform this cognitive task. As she travels northern

Europe with the rope-dancers, she believes that maps hold the key to her need to return home. But the protagonist in the story, Meister, notes the child's "incapacity . . . to represent anything . . . her body seems at variance with her mind" (quoted in Steedman 1995, 24). Steedman avers that Meister is pointing to a contradiction whereby Mignon finds extraordinary difficulty in controlling her body and yet at the same time can use it with great deliberation and precision. There is something wrong and disquieting about this child but Meister cannot find the words to express what he feels.

The important point about Steedman's (1995, 29) discussion of Mignon is that although the image derives from Goethe's text it resides remarkably intact in the cultural capital of nineteenth- and early twentieth-century Europe. She points out multifarious uses of the Mignon-figure in nineteenth-century literature and classical music. The broken body of a child-like figure is featured importantly in the monster in Mary Shelley's *Frankenstein* (1818). It is present in Dostoevsky's first sketches for *The Idiot* in the 1860s and in the song compositions of Schubert, Beethoven, Liszt, Gounod and Tchaikovsky. In Mignon's song "Kennst du das Land," for example, Steedman (1995, 29) points to a naïve and sentimental longing for home and the simple return to it. Mignon is travelling throughout *Meister* so the song is important not only for nineteenth-century Europe but because, from a contemporary reappraisal of its significance, it sets up a tension between mobile and sedentary identities that distinguish maleness from femaleness in the patriarchal order (cf. Aitken and Lukinbeal 1997). That the third verse of this song is interpreted to represent Mignon's castration further complicates the search for sexual meaning. Meaning is complexly represented in different ways in the many plays and stories that incorporated Mignonesque figures through the late nineteenth and early twentieth centuries. The androgynous, castrated and exotic Mignon figure of unspecified age is pervasive through the early part of this century to the extent that Margaret McMillan is not only able to use the term to political effect in her description of inner-city slum children in turn-of-the-century London but also in a way that unread working-class members of the British Labour Party could understand.

Not only is Mignon's sexuality uncertain, but her age is usually undetermined and may wander between infancy and adolescence in the same story. Sometimes she develops in stature and maturity as her sexuality is made more explicit but for the most part her uncertain age and shifting sexual status is an integral part of her story. This failure to place her on any kind of developmental map stands at odds with the evolutionary theories of the nineteenth century that presage the work of Piaget and Freud.

A contemporary evaluation of Mignon's narrative is that she cannot quite be placed and so she dies! The question that Steedman asks of our times is how to rescue her and find her a home:

The story of Mignon since 1899 is a story about the permeation of these cultural needs and desires: first to find the child, and then to give the child a home; and in these ways it is not surprising if we foget that Goethe gave his child a story once she was dead.

(Steedman 1995, 41–2)

As Steedman points out, Mignon's doom is not related to her development or sexuality but because of her strange position in relation to the story. The strangest part of Mignon's story is that the romance with childhood (and the placing of children) at least in the nineteenth century, is also about a romance with nation-building and the establishment of a global world order. Birth, death and mutilation come together in a myriad Mignonesque stories at precisely the same time that indices of infant mortality are statistically placing countries in the global north at the advanced pole of a developmental continuum and male circumcision becomes a familiar practice in these same countries. That childhood is saved by imagining the maimed, dislocated and broken bodies of child miners, textile workers and chimney sweeps is equally as important as its saving through educational and health legislation.[7]

The point that I draw from Steedman's (1995) interpretation of the Mignon story is that this strangely deformed child is created for the aesthetic needs of adults, that she is unable to locate herself on adult maps and, importantly, that she is a fulcrum around which sexual desire is gathered. But there is another part to Mignon's story that needs pointing out here. Specifically, it is important to look at how globalization, capital circulation and accumulation continually and profoundly transform the laboring body. The strange dislocations of Mignon's body are reflected in the warping of the bodies of children who labored in nineteenth-century mines and textile plants. Health and educational reform removed children from these contexts to better prepare them for a longer and more productive work life.

Children's bodies, technology and capitalism

Recent studies suggest that the interactions between children and technology need careful consideration of how the links and alliances between individuals, between and within communities and subcultures, and between bodies and machines are conceived (Bingham *et al.* 1999; Holloway *et al.* 2000). Felicity Callard (1998, 393) raises the point that some contemporary theorizing about the relations between bodies and technologies suggests "the newness of our 'current' condition has something to do with its difference from a previous world where, at least ostensibly, taxonomies were orderly and boundaries clear." The argument of some postmodern writers is that bodily fragmentation, dislocation and fluidity is novel and that at least part of its current transformation stems from material advances in modern telecommunications and hi-tech ergonomics. Callard raises two objections. First, it is unclear as to what is meant by fluidity and fragmentation and the level to which this fragmentation is said to occur in children is not well understood. Second, as the stories of Mignon and chimney sweeps suggest, the novelty of the modern fragmented child body is as much alleged as real. For surely, Callard argues, the fragmented body is as much a central figure in modernity as it is within postmodernity? It was only when the child laborer's body was torn and reconfigured at the face of coal mines or in chimneys that the division of labor, of bodies, is seen as characteristically capitalist. Callard cites the first volume of Marx's *Capital* (1976, 482) to elaborate the status of the workshop, now part of our cyber-connected private sphere:

> Unfitted by nature to make anything independently, the manufacturing worker develops his productive activity as an appendage of [the capitalist] workshop. . . . The process of selection starts in simple co-operation, where the capitalist represents to the individual workers the unity and the will of the whole body of social labor. It is developed in manufacture, which mutilates the worker, turning him into a fragment of himself. It is completed in large-scale industry, which makes science a potentiality for production which is distinct from labor and presses it into the services of capital.

The question that Callard raises when she alludes to Marx's notion of fragmented bodies is how are we to conceive the connections and differences between body–technology interactions in the world of early capitalist machine development, and those of twenty-first-century telecommunications? Her point is that a celebration of the fragmented subject through queer theory and post-humanist sensibilities in some senses plays into the machinations of capitalism, and therefore we need perhaps to detach Marx from his relegation to the master narrative and think more clearly of the implications of his work.

Resisting interiority

Callard (1998, 388) warns that in the rush to couch academic theorizing in metaphors of corporeality, perhaps certain aspects of the body are overdetermined. The other side of this critique is that perhaps some aspects are underutilized. Gregory (1994, 157–9), for example, is concerned that some postmodern writers such as Zukin and Soja lose site of Lefebvre's insistence on the body as a site of resistance. Virginia Blum and Heidi Nast (1996) elaborate on this suggestion, noting that although it is important, Lefebvre's notion of corporeality is not taken far enough because his subject is disembodiment and passively locates itself in some omnipresent, apolitical and reduced ideal. I argue here with Steedman that the reduced ideal that permeates work on subjectivity from Descartes through the functional morphologists to Lefebvre is that of interiority.

As a way of explaining some fundamental changes that he observed in nineteenth-century writing and practice, Foucault argued in *The Order of Things* (1970, 369) that European societies divested themselves of history. The new ''ordering'' was interiorized as people set about the task of discovering ''in the depths of [themselves] . . . a historicity that linked to them, essentially.'' Later, he wrote that it was not so much this new sense of self that is a development of the modern world but rather it is the location of the self that is new (Steedman 1995, 12). The new spatial and corporeal sense of this interiority is discussed quite copiously by geographers and other social scientists and I allude to some of this work at the beginning of this chapter (see also Pile and Thrift 1995), but links to how children and childhood are conceived remain elusive.

At the time that the process of interiorizing was going on, a large-scale reorganization of the way the natural world was perceived that began in the seventeenth century was coming to fruition. As I noted earlier, developments in the writing about scientific objects were changing, especially in plant physiology and ecology, to reflect a natural history of nature that mirrored developments in biology and anatomy. Foucault (1970) argues that human beings were being separated not only from the natural world but also from their bodies with separate narratives and namings where before they were described as together.[8]

Steedman (1995, 12–13) argues further that images of children, and, more generally, the idea of childhood, evolved at this time to express the depths of our inner selves. That these discursive changes occurred at the same time that psychoanalysis and studies of cognitive development were on the ascendant lends even more credence to the relationship between childhood and interiority:

> The idea of the child was the figure that provided the largest number of people living in the recent past of Western societies with the means for thinking about and creating a self: something grasped and understood: a shape, moving in the body . . . something inside: an interiority.
>
> (Steedman 1995, 20)

It may be argued, then, that the personal geography embodied in a child is used to represent human ''insideness'' to the extent that each individual has a history and a geography that promotes the development of an interiorized self.

This insideness is also, of course, about corporeality and the body. In an attempt to answer the question ''what is the body,'' Elizabeth Grosz suggests that:

Libidinal intensifications of bodily parts are *surface effects*, interactions occurring on the surface of the skin and various organs. These effects, however, are not simply superficial, for they generate an interior, an underlying depth, individuality, or consciousness. This *depth* is one of the distinguishing features marking out the modern, Western capitalist body from other kinds. Western body forms are considered expressions of an interior, of a subjectivity. . . . While social law is incarnate, "corporealized," correlatively, bodies are textualized, "read" by others as *expressive* of a subject's psychic interior.

(Grosz 1995, 34)

The point I want to end with is that this body, this interiorized self, is not just a human body, but a *child's* body. Through the beginning of this century – and particularly with the development of Freudian psychoanalysis – the interiorized self, understood to be the product of a personal and corporeal set of memories/geographies/histories seems most clearly articulated in the idea of "childhood" and the idea of "the child." The growth of popular contemporary psychoanalytic therapies that focus on the "inner child" help establish a discourse that articulates the contamination of children's worlds by abusive adults and the lingering shame that dysfunctional families engender in their children. Like Mignon, the inner child is lost, missing and abused, and in need of recovery, reparenting and reclaiming. Ivy (1995) interprets these images of contamination, shame and transgressed boundaries as a colonization of private, inner areas previously outside the realm of market structures and consumer compulsions. The increasing popularity of self-help books and therapies that focus on the inner child speaks to a larger social transformation wherein capitalist consumer culture targets interiority in sensationalized (missing children) and depoliticized (inner child therapies) ways. To the extent that dysfunction (particularly family dysfunction) is normalized (we all need to attend to the neglect of our inner child), when do Mignonesque deviations from some ideal form of childhood come to represent not just an odd difference from within a structured and coherent set of globalized images, but the signs of dissolution, reshaping and social transformation? The important point here is that children and their bodily experiences have not always and everywhere been used as emblems of the adult condition, though this appears to be the imaginative, psychoanalytic and cognitive legacy with which we operate at the beginning of the twenty-first century.

Notes

1 Marilyn Ivy (1995) links a growing American concern with missing children and child sexual abuse to increasingly popular "inner child" therapies in the last thirty years.

These therapies are directed towards people suffering from various addictions because their "inner child" has gone into hiding and is missing from their everyday adult consciousness. Important for what I want to say here, she argues that the popularity of "inner child" rhetoric signals a discursive sense of threat to older categories of identity, including that of the protected child. Market-driven consumer identities are reaching into our subjectivities and these changes are marked by depoliticized media terror tales about missing children and new therapeutic orientations.

2 Although Australian Aborigines have practiced totemic genital cutting for millennia, the oldest account of circumcision is in an Egyptian tomb dated around 2400 BC (Gollaher 2000, 1). From these Western origins in Egypt, it became an essentializing component of Judaism and Islam where the practice is used for bodily purification and as a rite-of-passage into manhood.

3 There was also a prevailing medical notion, that continues today in many professional and lay circles, that a foreskin is simply a flap of skin that serves no purpose. Countering this latter assertion are some contemporary medical studies suggesting that the inner skin of the prepuce is different from most of the rest of the body's skin because it is a "thicket of minute nerves" with a "complicated structure with specialized parts that serve different purposes" (Gollaher 2000, 121–2). Some conjecture that these purposes are integrally involved in pleasure but it is, of course, difficult to measure what pleasure may be lost when the prepuce is removed.

4 Modern secular attitudes revolving around the notion of purification from disease and healthcare are changing. A growing number of Western pediatricians now dispute the wisdom of operating on the genitals of healthy infants. Another reason for change is the recognition that for children and adults (if newborns could only speak!), circumcision is an excruciatingly painful surgery. Many proponents of the surgery in Europe and North America remain unconvinced, "likening the operation to a kind of vaccination that offered a lifetime of protection against cancer, urinary tract infections, sexually transmitted diseases, and even AIDS" (Gollaher 2000, xii).

5 Thomas Laqueur (1990) argues that for much of history, women were thought to have the same genitalia as men, the only difference being that theirs was inverted. For Laqueur, this provides an appropriate metaphor for the subsequent domination of women. From the perspective of circumcision, it suggests that the rationale for female cutting of genitals was similar to that of young males – control and punishment (cf. Gollaher 2000, 196).

6 It was not until the late nineteenth century that sexuality was defined as an essential core of individual adult identity (Foucault 1970) and, as I will elaborate more fully in the next chapter, it was not until the work of Freud that heterosexual meanings came to fix notions of femininity, masculinity and other differences.

7 The earliest English factory law (1802), that dealt with the health, safety, and morals of child textile workers, suggests that childhood is saved for children at this time. In the United States, early legislation was aimed at improving working conditions for children, but labor organizing was discouraged by the federal doctrine of conspiracy until this was superseded by state laws, beginning in 1842. Since then a wide range of labor legislation was enacted, including laws prohibiting hazardous occupations for women and children, and laws regulating interstate commerce to discourage sweatshop labor and child labor.

8 Indeed, Gollaher (2000) argues that it was a consideration of the foreskin as a separate structure in men's anatomy that propelled the medical practice of circumcision towards the end of the nineteenth century.

4

DRAPED OUTSIDE IN AND HUNG INSIDE OUT

Embodying sex and race

> The struggle is always inner, and is played out in the outer terrains.
>
> (Anzaldúa 1987, 87)

> The task of reality acceptance is never completed . . . no human being is free from the strain of relating inner and outer reality.
>
> (Winnicott 1971, 14)

A curious irony that derives from the nature mythologies discussed in Chapter 2 and the foundationalist principles of children's bodies discussed in Chapter 3 is the enduring idea that sexuality in children is repressed. Although we live in a post-Freudian age with an acceptance of the notion that children possess latent sexuality, it is none the less a common belief that children are beings who are not sexually active. All the supposedly repressed knowledge of childhood is loaded onto the early teenage years when bodily changes are most evident. A frighteningly disjunctive barrier separates the seeming sexual innocence of childhood and the ''public sexuality'' of adolescence.

Children's bodies are inviolable by convention and law to the extent that they cannot be interfered with sexually. Those adults who use children for sexual purposes violate such deeply held codes and mores that when caught they are often treated as less than human. Advocates of childhood as preparatory stages for adult life are often also adamant that it should be free from the burden of sexual relationships (Jordanova 1989, 16). Even those who advocate the legalization of sex between children and adults make a distinction between what they regard as 'consensual' and enforced – good and bad – sexual activity. The question of how these relationships are manipulated is equivocal and, as Stainton Rogers and Stainton Rogers (1999) point out, even when some arguments are dismissed as blatant appeals for legalizing pedophilia most attempts to permit some sort of sexual liaison ignore the power differentials between adults and children.

Power relationships are often directly proportional to age. For most countries, an "age of consent" sets an unequivocal border between those who know better and those whose bodies might be exploited. In the United States in 1861, the age of consent was raised from 10 to 12, in 1875 it was raised to 13 and in 1885 to 16. The variation between ages of consent amongst European Union countries is six years and in some countries the age differential between the partners is also taken into account. In others there are different ages of consent for heterosexual and homosexual sex (Stainton Rogers and Stainton Rogers 1999, 180). But there are fuzzy lines between the law and morality. For example, not everyone in Europe or North America would necessarily regard sex between a 15- and a 16-year-old as abuse and sexual play between prepubescent children is often dismissed as innocent exploration. Kincaid (1992, 70) argues that this tactic for defining childhood establishes at the center of the child a kind of innocence and purity, an absence of doing and an incapacity to make decisions. But discourses about innocence and purity turn on relations of power. The sexual abuse of children is seen as horrific because it is assumed to divest them of a form of innocence that is natural and unassuming. In some cases, innocence is used as a pretense to "protect" children from sexual knowledge that is deemed shocking or inappropriate.

The enduring story of Mignon, as highlighted at the end of the last chapter, suggests not so much innocence but an adult inability to place her (and the interior self), and culpability in the dislocation of her body and spirit. Her physically realigned body, her indeterminate sex and age, and her placelessness suggest an interesting set of metaphors for the positioning of contemporary childhood. What is clear from the preceding chapters is that the question of finding the place of childhood remains an enduring enigma precisely because of an adultist need to fix subjectivity. Gender is often one of the first things that children have to "get right," and it may be argued that the relative fixivity of other identities follows (Yelland and Grieshaber 1998, 2). In large part, sexual exploration is about exploring desire but it is also about the spaces of the body and, as such, it elaborates on "geographies closest in."

I want to draw from the previous chapter's exploration of specific constructions of children's bodies to begin my discussion of sexuality and difference with a clear understanding that the only universal stance I can make is that most children have bodies that grow and develop to various degrees. For me, the more appropriate questions revolve around the social construction of childhood and sexuality, and the way the adult world "infantisizes" to serve its own ends. From the critiques of Piagetian theory discussed in Chapter 2, it is clear that there are serious concerns about universal norms and developmental sequences. But we cannot

89

Figure 4.1 Gender trouble. This thought-bubble for a photograph I took at a local children's fun center was suggested by a colleague. It represents an adult interpretation of a "close-in" encounter in a bounded space. Who knows what was actually going on?

leave development there because Piaget was not the only influential developmental theorist of last century. The respect to which infantisization is a norm will depend upon the developmental theory in question and no set of theories position this as well as those that were articulated by Sigmund Freud and re-invented by Jacques Lacan.

The chapter begins with how Freud's ideas complicate the notion of sexual innocence but also fix sexuality as heteronormative. I then weave Lacan's linguistic turn into Freud's work with the caveat that he still maintained much of its patriarchal orientation. A group of feminist writers use the work of Freud and Lacan to suggest that the oedipal phase provides entry into a patriarchal culture that recognizes the construction of only two bodies – male and female (cf. Kristeva 1982; Irigaray 1985; Grosz 1994). I use this feminist critique to suggest that the twin notions of space and embodiment are important but neglected aspects of childhood that help elaborate sexuality and difference. In this latter discussion I draw heavily on the work of Heidi Nast (Blum and Nast 1996; Nast 2000) that tries to move beyond the articulation of sexual binaries. This, in turn, leads me to suggest that Winnicott's speculations on object relationships have strong claims, ones that he does not make, which focus on appreciation of difference and tolerance.

Innocence and the sexual child

As the most important twentieth-century figure to explore the *terra incognitae* of the sexual self, Freud generated diverse ideas about the formation and unstable character of the psyche. The very diversity of his work constitutes one difficulty for those who wish to use it, for although Freud's paradoxes may illuminate important and volatile issues, they can be difficult to reconcile in the process of theorization (Aitken and Herman 1997). That said, Flax (1990, 53), claims that ''Freud's writings on the constitution, limits, and powers of the self both challenge and reinforce Enlightenment views of humans as essentially rational beings.'' None the less, his deterministic view of repressive social relations is particularly problematic because it is inherently depoliticizing. This depoliticization stems directly from the basis of Freud's instinct theory and his tendency to consider gender to be categorical and unchangeable. Manifestations of gender roles are natural in the sense that variation or redefinition is seen by Freud as evidence of pathology.

The secret of human nature

> It is time we stopped imagining that the subject of sex tied the tongues of all Victorians, made them blush and stammer and stare at their shoes. It just as often brought a happy glint to the eye and inspired a poetry embarrassing perhaps only us.
>
> (Kincaid 1992, 134)

Freud focused on children's instincts as natural and innocent and he was at pains to point out that in children there exist forms of latent sexuality that must be contained and funneled in appropriate ways so that they can achieve a civil form of adulthood. Previous associations between changing bodies and growth, and between development and childhood were theorized in what emerged as psychoanalysis between 1890 and 1920.[1] Like Piaget, Freud adopted and adapted both recapitulationism and neo-Lamarckianism from Darwin's writings on human development (Davis and Wallbridge 1981). Morss (1990) argues that one impact of Darwin's work on Freud was to biologize childhood, emphasizing heritability rather than variation, and Burman (1994, 4) adds that the ''use of evolutionary assumptions to link the social to the biological provides a key cultural arena in which evolutionary and biologistic ideas are replayed and legitimized.''

91

But instinct and evolutionary theories alone provide inadequate backgrounds to explain the extraordinary innovation of Freud's ideas. The remarkable notion that sexual desire was repressed from childhood, that it comprised a large part of the unconsciousness and that it could be revealed through dreams was beyond anything being discussed in natural sciences at the time. David Bakan (1990) notes that there is very little in turn-of-the-century biology and physiography to suggest that the "secret" of human existence is sexual in nature.

It is reasonable to argue that the culturally constructed history and geography of Freud's ideas on the oedipal journey were fomented, at least in part, from the middle-class notion that sexuality was repressed in northern Europe and America during Victorian times, and this had a lot to do with the social construction of childhood. In his *History of Sexuality* (1980), Foucault notes that what appeared to be sexual repression and inhibition was actually a compulsion to create discourses about sexuality and children. The apparent fear and anxiety of Victorians around subjects such as children masturbating, for example, was translated into countless books and pamphlets on the topic.[2] Gollaher (2000, 100) notes that until the middle of the nineteenth century children's tendencies to play with their genitals were rarely remarked upon, but amidst a general shift in concern about promiscuity in cities, popular views of masturbation darkened. In the United States, the Comstock Act of 1873 expressed puritanical anxieties amongst the middle and upper classes about infant and child sexuality. Named for Anthony Comstock, a zealous crusader against what he considered to be obscenity, the act criminalized publication, distribution, and possession of information about or devices or medications for "unlawful" abortion or contraception. The act also banned distribution of such material through the mail and import of materials from abroad. Over the next quarter century, puritanical ideas and medical science proclaimed the evil of masturbation. Gollaher (2000, 101) quotes the popular 1896 monograph on masturbation of New York physician Joseph W. Howe who proclaimed that "pulmonary consumption, whose horrible ravages in Europe ought to give alarm to all governments, has drawn from this very source its fatal power." Gollaher (2000, 104) refers further to a paper read by J.B. Webster to the Ohio Pediatric Society profiling a typical masturbator as a 3-year-old boy, with "a scowl on his face . . . wearied and bloated . . . nervous and fretful, a poor eater and a very poor sleeper."[3]

The Foucauldian project is not only to argue that it is wrong to think children's sexuality had previously been passed over in silence, but also to put these discourses on sexuality in the context of power relations in society as a whole and work out their spatial effect. Power relations and what constitutes discourses of power are indelibly linked with surveillance and the control of bodies and spaces.

The end of innocence

David Archard argues that contemporary ideas of the sexual innocence of children are not only largely contradicted by the known facts but they also constitute Foucauldian discourses of power. Talk of innocence hinders the empowerment of children through limiting their access to knowledge and their attainment of awareness:

> Most worryingly, innocence itself can be a sexualized notion as applied to children. It connotes a purity, virginity, freshness and immaculateness which excites by the possibilities of possession and defilement. The child as innocent is in danger of being the idealized woman of certain male sexual desire – hairless, vulnerable, weak, dependent and uncorrupted. In sum the ideology of ''innocence'' may not protect children from sex. It may only expose them to a sexuality in the face of which innocence is debilitating.
>
> (Archard 1993, 40)

Moreover, Archard notes, explicit sexual education prior to the onset of puberty is viewed suspiciously because it may corrupt children with inappropriate knowledge.

What is deemed inappropriate? Thought to be preternatural in young children, explicit sexual knowledge is forbidden or couched in naturalistic terms. For example, children are often told that sex is nothing more than the mating of two animals thereby de-eroticizing the act. As Blum and Nast note (1996, 560), ''the apparent fixity of reproductive gender roles whereby mothering and fathering are construed as the only representational modes for sexual activity as well as the ultimate goal of the sexual relation . . . has the effect of de-eroticizing the sexual relation.'' Central to this view are naturalistic theorizations of hetero-sexuality and its attendant binary opposition between boys and girls. Although anyone can cross this division to which they are anatomically placed, to do so they must first line themselves up on one side or other. As Judith Butler (1990, 49) points out, ''lesbian sexuality . . . is a refusal of sexuality *per se* only because sexuality is presumed to be heterosexual.'' It is an either/or situation whereby gender is fixed through naturalized, heteronormative sex, and children are more often than not enculturated to accept the fix.

Getting it right

Freudian and Lacanian notions of identity are fixed around norms that prescribe developmental stages and hegemonic classes. But the shaping of environment and self occurs not only in psychological developmental processes that help form an individual, but also in the cultural, historical and geographical contexts. Certainly, at an early age, sexual, racial and class boundaries are constructed and maintained by children and caregivers. And, if these boundaries are ascribed as natural then young children may not be allowed to play with their identities because they have to get them right.

For Freud (1965), from as early as a few weeks of age, a beginning transformation occurs in an infant that precipitates the oedipal journey as it relates to the discovery of sexuality. Abnormal outcomes, such as sexually deviant adults, are explained by Freud as failures to move on successfully from the pre-oedipal or id-driven. Throughout his work, but particularly in *Civilization and Its Discontents* (1961a) and *The Ego and the Id* (1961b), Freud proposed a structure wherein the self is achieved through the delimitation of an external object or environment. This presupposition is developed in interesting ways in the early work of object relations theorists such as Melanie Klein and Donald W. Winnicott. I will argue later in the chapter that the writing on object relations by contemporary feminist and post-structural theorists such as Jane Flax and Debra Morris suggests an important unhinging of subject positioning. But to get to that place requires some consideration of Freudian, Lacanian, post-Freudian, post-Lacanian and post-structural thinking about sexuality, development, bodies and space.

The space of childhood proposed by Freud is not necessarily a *tabula rasa*, then, but it is a bounded space of innocence upon which a developmental transformation to adult knowledge is mapped. Like Piaget, Freud insists that the move from one stage to another, from pre-oedipal to post-oedipal, marks a normal unilinear progression from a simpler stage to one that is more complex. Moreover, the latter stage includes and comprehends the former as a component that is re-integrated at the higher level. The development of *self*, from basic Freudian theory, is an agonizing process of understanding what is left when the familiar object or person such as the mother is removed. Within this framework, original pre-oedipal wishes and demands become unsatisfiable and, while they continue to influence the formation of identity, are condensed and transformed such that they are safely confined to the unconscious. The transformation masks the powerlessness and dependency of the child, allowing the conscious mind to continue to function without crisis.

According to Freud, the way that an object/other is positioned determines the shape of the self. With time, then, a child learns to separate conceptually her- or himself from the environment with which he or she is a part. Lacan (1978), in his linguistic revisioning of Freudian theory, calls this phase the ''mirror stage'' and relates it to the development of language. In Lacan's account, the mirror stage, anticipated somewhere between 6 and 18 months of age, initiates the recognition of a distinct physical body, both in the person of the (m)other and in the infant.

Prior to this stage, the new-born infant does not distinguish her- or himself from the environment (and especially the mother). Imperfect mirroring by the mother, which is an inevitable consequence of the growth of awareness, forces the infant to recognize dependence and induces a crisis of self-esteem. At this point, it is important to note Lacanian and Freudian agreement that separation is a process whereby children learn to become cultural, monadic selves and to recognize the difference between themselves and "others." Conceptualizing this distancing of the child from their environment, also known as the division of "self" from "other" or the separation of "subject" from "object," has played a crucial role in Western thought. Marwyn Samuels (1978) and Steve Pile (1993), two geographers with widely different philosophical perspectives and writing at different times, agree that this distancing between a person and their environment, this attempt to discover "the fragmented self," is not only fundamentally spatial and geographic, but it is also an essential characteristic of the human condition.

Anti-Freudian perspectives dominated developmental theories from the 1950s onwards. Most, like those emanating from the work of Piaget and Erikson, simply avoided Freud's focus on the fragmented self. A notable exception is prescribed by the work of John Bowlby (1951, 1988) who took on Freud from a biological perspective by arguing that attachment rather than separation is the key to understanding human development. Bowlby developed a series of psychoanalytic theories that were anti-Freudian in the sense that they offered a biological rather than a sexual motive for attachment to the mother. He saw the new-born infant as an unsociable creature whose proper socialization and development necessitate a mysterious process of attachment to the mother. Ethology provided an inspiration for Bowlby's work and, from this base in the scientific study of animal behavior, he argued that attachment has little to do with sexuality. Anguished by observing the extreme effects of separations he observed in various species of animals as well as people, Bowlby (1988) notes that the person who matters most is neither the feeder nor the sexual partner, but the one who stays near and offers protection against danger, the unknown and loneliness. Bowlby joined with developmental psychologist, Mary Ainsworth, to experiment with attachment patterns. Through observations of mother–child interactions around the globe, Ainsworth (1972) demonstrated that the child's exploration of the environment beyond the mother is a negotiated developmental process for both the parent and the child but that attachments are none the less enduring. Like Bowlby, she claimed that mother–child attachments are natural and instinctive, and that their disruption can cause permanent psychological damage.

Many Freudian and post-Freudian theorists speculate upon the sexual nature of attachment and its relationship to how the child is gendered although many feminists are rightly suspicious of talk about a natural or seemingly sacred bond between mother and child. None the less, some are convinced by the Lacanian notion that the dynamics of desire in the mother–child relationships underscore the mother's desiring subjectivity as well as her pivotal role in producing a desiring subject, the child. This desire focuses on the phallus that comes to signify loss, division and a fragmented identity that can only be re-forged through connection with the maternal. The child contrives to recover its place in the natural/maternal and to overcome its loss by attaining what the mother wants, namely the phallus. For Lacan, this connection can only be pursued indirectly through the order of language (Blum and Nast 1996, 570–1). Alternatively, Freudian theorists see the development of child–mother separation solely in terms of sexuality, and divisions and losses around the recognition of the power of the mother figure. That said, it is misleading to suggest that either Freudian or Lacanian theorists are readily prepared to see children in sexual terms because childhood is still represented as a period of asexual innocence and this innocence is one of the aspects of loss. Moreover, and importantly, when desire in the child evolves it is presumed to revolve around heterosexuality as a pre-discursive norm. In reference to the onset of the oedipal journey, Judith Butler (1990, 60) avers "to what extent do we read the desire for the father as evidence of a feminine disposition only because we begin, despite the postulation of primary bisexuality, with a heterosexual matrix for desire?"

The symbolic order and nature

From Lacan's (1978) mapping of textual and symbolic dimensions onto Freud's theories of the oedipal complex, there is the suggestion that language provides a basic structure for a child's transformed understanding of self and the world. Lacan's locating of identity in language stands in opposition to the biologism of Bowlby, Ainsworth and others that pervaded developmental theory in the 1960s and 1970s. Yet, as Blum and Nast (1996, 569) note, Lacan's anti-biologism leads him to locate subjectivity entirely in language which obviates bodies and spaces as effects and containers respectively. The "real" – some form of pre-discursive natural space or primordial connection with mother/nature – is lost for Lacan during the crisis of loss of omnipotence that accompanies the mirror stage. This crisis leaves the individual unable to recapture the real except through fantasies that are constructed in and mediated by language. Language fulfills the function Freud attributed to the condensation and displacement of original desires during the "mirror stage, which can be viewed as a chain of signifiers (having primacy over the signified) that connect the speakers to the real" (Walkerdine 1988, 191). Although they note that Lacan's real is anything *but* a naïve notion of "reality," Blum and Nast (1996, 561) point out that there are times when this order resembles "nature" in the way that it both opposes culture and is connotatively linked to the maternal. Importantly, for Lacan, the real is regulated and controlled not only by language, but also (occupying the same position as language) by the symbolic presence of a third party, the father (Bondi 1996). If children do not separate from the real they are subject to the invidious and psychotic lure of the maternal. Abiding by paternal laws (the Law-of-the-Father) that ensure, for example, the prohibition of incest and the consequent punishment by castration, ensures the developmental transformation of a child into the symbolic chain of signification. For Lacan, difference – what Blum and Nast (1996, 561) call *alterity* – is founded in renouncing the mother's body.

Many feminists point out that the writings of Freud and Lacan are androcentric and misogynistic, positioning feminine sexuality as the dark continent, the forbidden place of the oedipal myth, and women as deviations from the male norm (Grosz 1990). Alternatively, some feminists find promise in Lacan's account of the formation of subjectivity because ''his problematization of what psychology usually takes for granted enables us to shift the axis through which we pose our questions'' (Urwin 1984, 275; but see also Grosz 1990). Constituting the critical process through which subjectivity is produced, Lacan suggests that language acquisition provides continuity in society since the complex of signifiers that is transferred is, in effect, reproductive. This proves an intriguing platform from which to address the persistence of dominant and repressive systems of knowledge, but Lacan's account of the use of symbols is problematic for some feminists because it grows out of an essentialist and apolitical view of human nature. For Freud, narcissism is positioned as an irreducible aspect of human nature, the quality of which necessitates self-identification and the splitting of the subject. For Lacan, language rescues the ego in a way that is preferable to direct submission and dependency (Bondi 1996). None the less, individuals are henceforth split by the need to voice their desires to an other who is independent. To Lacan, language's chain of signification is rooted in the narcissistic stage and operates as an independent force. He argues that language provides symbols that pre-exist the individual and, thus, structure object relationships that are formulated throughout life, even though the individual labors under the illusion that he creates his own symbols. In Lacan's framework, the symbolic is patriarchal and the social order is regulated by an abstract father whose position of power is inviolable. Individuals cannot appropriate or manipulate symbols but are confined by a pre-existing system of signification. This denial of agency leaves culture as a static entity and disables any opportunity for it to be discursively reproduced. So, although the transference of signifiers through language acquisition is the basis of reproduction, what is reproduced is fixed by heteronormative assumptions.

Alterity, space and embodiment

Feminist and post-structural critiques of Piaget's child-centered position, Freud's oedipal journey and Lacan's mirror stage point out that they are forms of knowledge that are instrumental and masculinist to the extent that their social and cultural construction is hidden in claims of universal truth. Sexual and racial identities are either not considered at all, or are thought to be irrelevant or subsidiary to the 'true' process of development. The 'plasticity' of sexuality and human development provides a focus for some theorists who down-play the universality of psychological and social phenomena and focus rather on their creation as personal and cultural products (Henriques *et al.* 1984; Walkerdine 1988; Brandtstädter 1990). For post-structuralists there is no "true nature of the child." Adherents to this perspective are not comfortable placing children into particular developmental or sexual stages, although important characteristics of development may be evident at certain ages and under certain laboratory conditions. Post-structural perspectives de-center encompassing scientific theories and meta-narratives in favor of discourses which take into account the social construction of their own knowledge base.

For Henri Lefebvre (1991), structural theories – whether focused on development (Piaget, Lacan), cognition (Levi-Strauss) or language (Saussure, Barthes) – tend to universalize, systematize and take for granted bodies and space. By so doing, they eliminate possibilities for creative and subversive acts as well as structural transformations (Blum and Nast 1996, 559). As I pointed out in Chapter 2, space for many of these theorists is a passive container for, or register of, meaning that is actively generated by cognition and/or language. For Lacan, in particular, the privilege of the image (or the image of an image in the mirror stage) presupposes the visual over the spatial, and the spectacle over the corporeal. Blum and Nast point out that Lacan's body is reduced to two dimensions with a possibility of a third dimension in the phallus, which is the dimension that founds and mediates alterity. As a signifier without a signified, the phallus produces and sustains meaning and identity without implicating or locating itself (the view from nowhere).

In pointing out the debt spatial theorist Lefebvre (1991) owes to Lacanian psychoanalytic theory, Blum and Nast (1996) argue that both theorists focus on gender construction as *the* fundamental social process through which alterity is achieved. Nast (2000) then takes some of these arguments further to suggest an important link for theorizing racial differentiation. What may be possible is the development of Nast's work to suggest a non-heterosexual spatial domain for the embodiment of children's identity. Blum and Nast argue that a non-heterosexual spatial domain is a set of constructs that Lefebvre points to, but never develops. This omission works towards my argument in favor of the development of contemporary, post-structural readings of object relationships that I believe help focus a psychoanalytic lens on notions of spatialities of difference that are not only tolerant but liberatory. I'll come back to this at the end of the chapter.

According to Blum and Nast, what distinguishes Lefebvrian from Lacanian thought is its potential for transcending phallogocentristic and heteronormative fixing of identity. They note, however, that this remains only a potential because Lefebvre does not distance himself sufficiently from Lacan's phallogocentricism to engage non-heterosexual forms of identity formation. Blum and Nast note that both Lacan and Lefebvre locate the origin of the subject in a prediscursive maternal realm, and it is out of separation from this realm that difference and subjectivity emerge. The Lacanian mirror stage is a complex negotiation whereby the child locates herself through an image that is outside her own subjectivity but with which she identifies. A physical internalization of this image founds the child's subjectivity. As Blum and Nast (1996, 564) point out, "subjectivity is spatially and ontologically *decentered;* the subject is shaped literally from the *outside in*." Lefebvre's problem with this account is that the subject is disembodied and passively locates itself in some omnipresent, apolitical, two-dimensional mirror. But rather than discard this metaphor, Blum and Nast point to a critical passage in Lefebvre's work, that I expand upon here, where he weaves the mirror metaphor with his concept of the mirage (as something more fantasmatic than the glass) to produce a space that is at once active and multi-dimensional:

> The mirage effects . . . cannot be reduced solely to the surprise of the Ego contemplating itself in the glass, and either discovering itself or slipping into narcissism. The power of a landscape does not derive from the fact that it offers itself as spectacle, but rather from the fact that, as mirror and mirage, it presents any susceptible viewer with an image at once true and false of a creative capacity which the subject (or Ego) is able, during a moment of marvelous self-deception, to claim as its own. . . . Whence the archetypical touristic delusion of being a participant in [a landscape], and of understanding it completely, even although the tourist merely passes through a country or countryside and absorbs its image in a quite passive way.
>
> (Lefebvre 1991, 189)

The creation of subjectivity for Lefebvre occurs with the projection of the entire material world fantasmatically onto the ego while, at the same time, the specular introjection of the landscape shores up the ego's sense of power and coherence (Blum and Nast 1996, 567). Alterity is founded in a movement from natural space to *absolute* space wherein the material world complexly takes the place of

99

Lacan's mirror. Blum and Nast trace Lefebvre's historical justification for the transformative role of space on subjectivity since before Greek times, but what is ultimately important about the concept of absolute space in modern times is that it comprises a patriarchal understanding of roles and power relations. Lacan's visual and linguistic reductionisms about the subject deflect the mirage of a larger material world and, as a consequence, are unable to theorize how gendered constructions are served politically. The tourist passing through a landscape is a passive, disembodied, bourgeois figure with the cultural and financial means to reduce the world to an image or a dialogue. Lefebvre does not deny the power of Lacan's mirror, but rather he constructs it as part of the process of visual signification endemic to commodification and capitalism and, as such, the mirror becomes yet another face of patriarchy (cf. Pile 1996). Of course, Lefebvre's mirage is also a face of patriarchy but unlike the Lacanian subject, the Lefebvrian subject is always bodily, spatially and politically embedded in a material order. None the less, Blum and Nast (1996, 568) criticize Lefebvre's reworking of the mirror for not going far enough. He establishes the embedded material power of patriarchal order without any discussion of the struggles of those who are disempowered by gender, sexuality, race/ethnicity, abilities and age. Trans-formations of that material order and how it circumscribes the struggles of young people are the subject of the next two chapters, but before I can pick up that discussion I need to move identity theory beyond its stultifying roots. Some of this may be possible by recognizing how much these roots are influenced by Cartesian logic.

At one level, Lefebvre's "social body," like the body of Mignon, is devastated and broken by a division of labor that has very little to do with material or natural spaces. At another level, he theorizes a space that is undifferentiated from the body. His "spatial body" is produced and conceived by space and is immediately subject to the determinants of that space:

> Symmetries, interactions and reciprocal actions, axes and planes, centres and peripheries, and concrete (spatio-temporal) oppositions. The materiality of this body is attributable neither to a consolidation of parts of space into an apparatus, nor to a nature unaffected by space which is yet somehow able to distribute itself through space and so occupy it. Rather, the spatial body's material character derives from space, from the energy that is deployed and put to use there.
>
> (Lefebvre 1991, 195)

The social body is first recognizable with the separation of productive activities (and a consequent division of labor) from reproductive activities, both activities having formerly been located in the household (Lefebvre 1991, 49). This body is later abstracted and transformed through capital with the commodification of labor and its location in an emerging private sphere. Progressive abstraction of spatiality and labor continued with the embracing of Euclidean logic to demarcate space and situate the world in absolute terms. The multidimensional performance of lived experience could be more easily charted and controlled on a two-dimensional cartographic representation with a coordinate grid to fix a subject's absolute location. So, as I argued in Chapter 2, why not also fix a child's development through cartographic means? Kathleen Kirby (1996, 45) argues that "otherness" (of women, children, minorities and so forth) was erased by a mapping that applied its own culturally specific standards. Subjects, like places on a map, were homogenized in favor of a generic iconography to the extent that any social policy based on humanism became insensitive to people's varying needs.

The important point that distinguishes Lefebvre's work is his insistence that although the flattening of the lived world into a Euclidean space is understood textually and linguistically, there is something beyond these two dimensions that is doing the flattening. Blum and Nast (1996, 573) argue that this force resembles Lacan's phallus:

> This might be the upright pen (which writes down the quotas and issues the pink slips) as well as repressive state apparatuses (which materially enforce such symbolic acts). In other words, for Lefebvre, the phallus is the godlike verticality of force and agency that capitalism assumes in order to signify the world in ways that further its own goals. . . . Those not allowed such verticality, those who are commodified and scripted into place, make up the not-vertical, the oppressed.

Unlike Lacan's apolitical (signifier without a signified) phallus, Lefebvre's version is corporeal, spatial and political. Although they laud his attempt to bring together "intrapsychic, spatial and corporeal aspects of subjectivity," Blum and Nast (1996, 568) are none the less critical that Lefebvre's account of history is still premised upon the irreducible gendered Cartesian mind/body split. They point out that his historiography of heterosexuality turns on an active–passive binary wherein materiality inscribes the body but he only considers those inscriptions which are coded masculine whereby feminized and racialized bodies necessarily become invisible. Masculine agency is privileged through its emphasis of masculine spaces while more mutable and feminized socio-spatial practices or hidden racialized struggles are ignored. The hegemony of Lacan's phallus reemerges in Lefebvre's single (male) agent (Blum and Nast, 1996 577). What this account misses is an engagement with writing about bodies that resist and transgress the dominant male/female coding of phallus/lack (e.g. Kristeva 1982; Butler

1990, 1993; Iragaray 1993; Grosz 1994). For, as Blum and Nast (1996, 578) aver, "just as Lefebvre posits that we always bodily and spatially exceed the surface of the mirror, so too we continually exceed the disciplining patriarchal codes of contemporary social orders."

The problem with Lefebvre's construction of identity in space and time is that it too is developmental because it casts the time of liberation and radically different ways of knowing into a revolutionary future that misses the point that different forms of corporeal and spatial identity have always occurred. Capitalism and patriarchy may be perceived as hegemonic but they are also partial. Therein lies the possibility of transgression.

Transgressive and geographically embedded bodies

Some academic feminists who write about the body offer different sets of metaphors for thinking a way out of dualisms. Elizabeth Grosz (1994), for example, uses Alphonso Lingis' metaphor of the Moebius strip to help her understand the erotic interdependencies of the interior and exterior. Moving along the outer surface of the Moebius, it folds over into the inner and moving along the inner surface eventually brings the traveler back to the outer:

> Lingis seeks to evoke, to replay in words, the intensities that charge all erotic encounters, whether the amorous relations of the carpenter to wood and tools, that attachment of the sadist to the whip, the liaison of breast and mouth, lips and tongue. There must be some coming together of disparate surfaces; the point of conjunction of two or more surfaces produces an intensification of both.
>
> (Grosz 1994, 197–8)

According to Grosz, just as the inner mind inscribes boundaries, units and functions on the body so the cultural capital inscribed on the surface of the body creates the feeling of inner depth, unity and identity in what we experience as the self. She uses the Moebius metaphor to argue that the point of origin for mind and body is the body's surface.

Alternatively, Judith Butler (1990, 1993) tackles the male/female binary theatrically by suggesting we attend to the performance of a fluid sexuality within and through space and time. She argues that a focus of this kind enables a cogent appraisal of the relations between bodies and space, or between bodies and the objects and environments in which and with which they move and interact. In a sophisticated critique of heteronormativity in Freudian and Lacanian psychoanalytic theories she broadens understanding of the political by opening up naturalized assumptions that perpetuate Cartesian logic and a gender/sex dichotomy. Butler (1990) uses Foucauldian genealogical methods that critically examine and disrupt the production of hegemonic knowledge and power to show how psychoanalytic theories naturalize heterosexual desires and place them outside the realm of the political. She argues that feminist distinctions between material bodies (sex) and social bodies (gender) fail to recognize how an anatomically derived male/female distinction is produced through discourses of gender. Instead, famously, Butler (1993) advocates theorizing gender and sexuality (or any identity) as performativity. As summarized by Lise Nelson (1999, 337), perfomativity ''recognizes that 'the subject' is *constituted* through matrices of power/discourse, matrices that are continually reproduced through processes of re-signification, or repetition of hegemonic gendered (racialized, sexualized) discourses.'' This articulation of space and bodies is important because it enables a move beyond Freud and Lacan's figural space of symbolic order through which subjectivity emerges to material spaces that are marked by resistance to exclusionary politics and heterosexual imperatives (Callard 1998, 390). Children and adults alike engage themselves and the world through performances that may be thought of as illusions created through particular events in space and time, which are repeated to the extent that they become part of the self. Butler's (1993, 2) notion of gender performativity as ''a re-iterative and citational practice by which discourse produces the effects that it names'' is insightful for my purposes because it suggests that children's experiences are embodied not only by themselves but also, and particularly, by adult and societal norms.

The performative nature of sexual identity, argues Elizabeth Gagen, ''is particularly fruitful with regard to children'' because

> in spaces where learning is the predominant enterprise, adults' efforts to induce particular performances confirms, first, that the aim of such practices is to acculturate children towards social norms, and second, that the alignment of certain gender performances with certain sexualized bodies is not inevitable, nor natural, but closely managed.
>
> (Gagen 2000b, 214)

In a study of the early children's playgrounds in the United States that I will discuss more fully as a material transformation in the next chapter, Gagen notes how certain performances were upheld and directed towards specific socio-spatial consequences. Playgrounds for older boys were devoted solely to sports such as baseball where forward individuals learnt how to perform heroically, to obey authority and be loyal on a team to the point of personal sacrifice. Consequently, they were prepared for civic life and war. Girls, in contrast, were taught quiet non-competitive activities such as sewing or knitting. Consequently, they were prepared for patriotism and domestic life. These, mostly immigrant, children were taught cleanliness, and the girls in particular were taught appropriate body comportment. Gagen (2000b, 220) argues that playground organizers endorsed Butler's notion of an "expressive model" of gender signification which legitimized the logical separation of boys and girls. By so doing, the organizers felt that their young charges were able to express a new-found, truer identity. The playground organizers concealed the flexible and performative nature of gender by institutional commitment to expressive (heterosexual and national) notions of identity that were mapped onto the children's activities.

By outlining how different psychic processes of identification disrupt any smooth and precise demarcation of territorial/national based identity politics, it is impossible to highlight spaces such as playgrounds as neutral boiler-plates that allow bodies to be read unproblematically. Bodies, and particularly the placing of bodies in space, matter, but it may be argued that linguistic theories such as Butler's performativity are too closely tied to Freud and Lacan to enable a liberatory understanding of the relations between identities and bodies. In her critique of performativity, Nelson (1999, 339) argues that while it is important to recognize that foundational norms within enunciations of identity are problematic, and while it is important to understand how the "compulsion to repeat" (Butler 1990, 145) re-signifies dominant discourses, the notion of performativity none the less forecloses inquiry into why and how particular identities emerge, their effects in time and space, and the role of subjects in accommodating or resisting fixed subject positions. These conceptualizations reproduce a semi-Cartesian dislocation from geography, history, location, context and anti-imperialist struggle. Nelson is concerned about how dominant discourses are subverted through the process of repetition. How, precisely, does social and political change occur? If repetition is regulated by dominant discourses, then change is possible only with displacement or slippage within the process of repetition. For Butler (1990, 28), change occurs with the spontaneous emergence of that which is repressed (e.g. nonbinary sexuality), but Nelson points out that this suggests a process or series of convergences (spontaneity) that operate autonomously from the subject (if not completely outside of it) because the slippages are not conscious or intentional. Nelson is concerned that although questions of agency, knowledge and participation – conscious subversion and/or appropriation of dominant discourses – in place and over time are beyond the parameters of Butler's analysis, they are the kinds of questions asked by geographers. Butler leads in an understanding that agency is not autonomous nor is identity fixed, but she sidesteps the issue of critical consciousness:

> Her theory of performativity treats any enunciation of identity as *necessarily* fixed, and any notion of agency as *necessarily* one that implies an autonomous, masterful subject. In other words, she runs into a

bind because she only conceives of conscious agency as stemming from an *autonomous* (pre-discursive) subject.

(Nelson 1999, 340)

Nelson argues that retheorizing identity and subjectivity as not fixed and exhaustive but changing and contested over time and space does not foreclose the idea of a conscious, thinking but not necessarily autonomous subject. Identity can be constituted through hegemonic discourses of race, sexuality, age and gender while subjects can remain critical of, and subversive around, the process. Nelson's enduring point is that as a textual theory, the limits of performativity emerge most starkly when identity is explored within particular geographical and historical contexts. She suggests that one way to reconceptualize agency without defining change as random and spontaneous or reifying subjectivity as autonomous is to think about geographical embeddedness. This kind of embeddedness suggests that human subjects ''do identities in much more complex ways than performativity allows.'' It is a negotiation and a struggle that is at times conscious ''but it is never *transparent* because it is always inflected by the unconscious, by repressed desire and difference'' (Nelson 1999, 347). Nast (2000) theorizes that this inflection is poignantly demonstrated by violence on colored bodies and colonization of their space.

Racism, infantilization and the political unconscious

In work that links racism to what she calls the racist oedipal family, Nast (2000) argues that ''race'' is inherent in modern conceptions of heterosexuality that construct a normative family quadrad comprising Mother, Father, Son and the Repressed (the Wild or Bestial). The Mother–Father–Son triad is encoded as white while the repressed Bestial – poignantly highlighted by the Wilding of Central Park described in Chapter 2 – is colored (but also young, virile and threatening). She argues that the oedipal and the pre-linguistic/linguistic split should not only be jettisoned, but that also we should scrutinize the ways interiorized unconscious anxieties register within exteriorized landscapes and how an understanding of these helps expose the political unconscious working at numerous geographical scales.

As indicated at the beginning of the chapter, Freud's work was influenced by functionalist and Darwinian theories, but Nast argues that it was also shaped by colonialism. Freud's work coincided with the onset of systematic African conquests following the 1885 Berlin Conference that partitioned the continent amongst European powers. Victorian ideals of racial whiteness are expressed, for example, in Freud's often remarked-upon conflation of the feminine psyche with Africa-as-the-Dark-Continent (Nast 2000, 224). What remains under-theorized in Freud's work is the socio-spatial violence, desire and repression that accompanied colonialism:

> The psyche was, in this sense, an interiorized repository within which violent acts and desires of colonization were secreted or made *legitimately secret* and *unspeakable* . . . the memories and actions associated with colonial violences were incorporated into the body-space of the "psyche," and unconscious domain outside language . . . [and] certain "unconscious" colonial violences were sexualized . . . as a libidinous foil against which the white oedipal family anxiously defended itself.
>
> (Nast 2000, 215)

Because it is outside of language, this embodied "discourse" also remains hidden from the scrutiny of Lacanian analysis. Indeed, Nast argues that most twentieth-century psychoanalysis is unwittingly formed by and instrumental to "white" colonial desires.

Nast attempts to access the repression of colonial violence inherent in modern cultural vitality and racism's immanence in the oedipal family by spatially reworking Fredric Jameson's (1984) notion of the political unconscious. Jameson's concern was to develop a form of Marxist analysis that respected and utilized textual and cultural differences rather than collapsing them into an undifferentiated reflection of class. For Jameson, texts and cultural meanings are, at their most fundamental level, political fantasies that articulate both the actual and potential social relations that constitute individuals within a specific political economy. The political unconscious is a means by which the fantasy may be addressed and Nast argues that his notion of the *imaginary–symbolic* is a powerful tool for interrogating nineteenth- and twentieth-century racism. In later work, Jameson (1992) argues that the space of the political unconscious may be "cognitively mapped" but his use of Kevin Lynch's (1960) conceptualization of *cognition* and *mapping* constrains insight to the instrumental and the Cartesian respectively. Nast (2000, 222) takes Jameson's work away from these and other structural techniques to show how binarisms point to dialectics of historical tensions and change. She argues, importantly, that although the psyche refers to an "inner" imaginary–symbolic space of unresolved conflict it is created and internally differentiated through "outer" spaces of struggle.

The spatial struggles Nast describes are related to colonizations that are constituted in large part through the imaginary–symbolic political fantasy of a racist oedipal. Nast constructs psychical interiorities not only as "real" (embodied spatial effects) but also as embodied geographical effects of colonial conquest. But in the practice of psychoanalysis, race is not theorized as part of unconscious desires and anxieties and, as such, Nast (2000, 218) argues that psychoanalysis is complicit in reproducing a white supremacist oedipal family. Her arguments dovetail with the larger debate on the construction of blackness in the nineteenth and twentieth centuries as bestial, incestuous, uncivilized, uncontrollable or evil. Nast uses this debate to propel the notion of racist familial constructions and socio-spatial oppressions. From the symbolic order of the oedipal family, the white mother is pure and is constituted as needing protection – unevenly in space and time, both bodily and geographically – from incestuous blackness and nonpurity.

Nast outlines a number of stories suggesting ways the psyche is linked to racist desires that produce an oedipal order wherein part of the family quadrad is constituted as colored and wild, and controlled through a symbolic imaginary of infantilization. Citing Anne McClintock's (1995) work, for example, she notes that industrialization and the nuclear family emerged at the same time as colonization in Asia and Africa, suggesting (at least in the case of Britain) a link to a nostalgic desire to rescue the remnants of a declining monarchical order with colonial royal subjects who were structurally infantilized through ''race.'' Other examples include the emasculation of black men as ''boys'' in the United States South and the denial of their status as fathers (often the white plantation owner took a paternal role over black children). After the Civil War, what had begun as the libinization of slave bodies and places in the colonial and plantation context became castration and death at the hand of the white iconic father (Nast 2000, 224). Centuries of physical and psychical authority over black bodies came to an end and white supremacist law and order was severely threatened because the black boys could now theoretically vie for paternal status. The key to Nast's (2000, 226) speculations about the racist oedipal ''is that white fears could not be spoken because they were so psychically submerged and interwoven through a labyrinthian maze of familial desire and political-economic and symbolic necessity.'' White rage is never articulated in words but, rather, it is spoken – past and present – through violent action and imaginary symbolic stories. These stories inculcate textual representations including, as perhaps the best example, D.W. Griffith's classic, racist silent movie, *The Birth of a Nation* (1915) where all black people (actually whites in black-face) are represented as child-like buffoons who are seemingly incapable of assuming adult responsibilities. In another example – one that clearly anchors her theories of spatial effect – Nast discusses the protracted socio-spatial actions of the University of Chicago in the face of substantial industrial related black in-migration to neighborhoods around the campus. The actions began in 1892, almost at the same time that the campus was formally inaugurated in the exclusive white South Side suburb of Hyde Park, when both the University and Hyde Park supported restrictive covenants excluding blacks from residing in the area. Other gambits included the use of Chicago's famous elevated railway as a barrier between white and black areas, creating a "black belt" from 1900 to 1930. Following the outlawing of racial covenants in 1948, massive urban renewal at the behest of the University resulted in undesirable buildings being torn down and rents increasing dramatically to "maintain the white population": "Using the skills and efforts of numerous politically connected and/or wealthy alumni and trustees, the University managed to make Hyde Park a showcase of the first national center-piece of the comprehensive [federal] urban renewal plan" (Nast 2000, 239). Grassroots support for the University's plans was cultivated through paternalistic manipulation of fears of black rapists. Nast characterizes these latter actions as oedipal hysteria, citing a frenzy by the authorities and the media after a series

Figure 4.2 The political unconscious. Norrie shows his distaste for and resistance to patriarchal oedipal racism. This picture is, of course, reproduced here completely out of context. Part of Norrie's story, and my relations to it, may be found in Aitken (2001)

of alleged rapes of white women (associated with the University) by black men. Eventually, the "boys" in neighborhoods surrounding the University were symbolically castrated by the much lauded urban renewal plan, and left spatially stranded and neutralized by a massive surveillant police force (Nast 2000, 242).

What Nast's account points to is an engagement with writing about bodies that resists and transgresses the dominant male/female coding of phallus/lack and the racist oedipal (e.g. Kristeva 1982; Butler 1990, 1993; Iragaray 1993; Grosz 1994). She achieves a re-theorizing of the pre-linguistic as a spatial effect of the psyche – both past and present – that is mediated through cultural belief in the oedipal. As Nast argues elsewhere children, families, communities and nations need to be made places of scrutiny *through the body* so as to continually undo repressive and violent sexism and racism (Nast 1998, see also Nast and Pile 1998). Nast's work points to the ways sex and race are embodied and embedded in dominant discourses of power, and what is of particular interest is the way these discourses infantilize in order to disempower. I spend some time with these ideas here to highlight what are clearly important links between the nature of childhood, corporeality, race and sex. The next step is to suggest transgressions, resistances and other routes out of this Cartesian miasma.

Borderland bodies and resistance

Jacquelyn Zita (1998, 167) argues that Gloria Anzaldúa's mythopoetic work is deeply contextual, historically and geographically located, and decidedly raced and sexed. Anzaldúa writes against Cartesian bodies from where she lives as a mestiza lesbian in Cargill Texas, at the southern edge of the global north. Whereas biologists from the nineteenth century onwards believed that all the activities of the body could be explained with an understanding of its material compositions and the interaction of its parts at the physio-chemical level, Anzaldúa suggests that the body be viewed "*as activity* in resistance, survival, and historical transformation." Anzaldúan bodies, unlike Cartesian bodies, break into geography, ethnicity, memory, political struggle, and ancient practices of the flesh. Her writing in *Borderlands/La Frontera* disassembles Cartesian writing practices by focusing on what is meant by the body and the mind and, importantly, their relationship to Anglo-American colonization and commodification of the body's matter. Zita traces the contours of the "body as activity" in Anzaldúa's text to the extent that it becomes resistance, survival and a geographical transformation.

For Anzaldúa (1987, 87), "the struggle is always inner, and is played out in the outer terrains," through open wounds, faces, skins, bones and blood. Images, words, stories – the stuff of research – are also corporeal and they are transformative because they fold back onto the body and remake the soul. The power is not in the body as sheer instrumental mechanism but through its metaphysical ability to deconstruct the polarities of dualism by physically blending and kneading (cf. Zita 1998, 179 and 180). A new borderland political consciousness is experienced by Anzaldúa by writing through and in the body. It is a consciousness that emerges through a body that is local, specific and living: a view from somewhere. Importantly, she gives the concrete events of daily, material life a space for larger reflective and critical knowledge that does not require the disappearance of the thinker's (researcher/adult/child) body.

Objects of inquiry

Even for very young children, desire and conscious action are mediated, encumbered and coddled by constitutive discourses. A situated and embedded approach to subjectivity and identity is one that appreciates intersubjectivity, self-reflexivity and knowledge production. It is one that opens up what Cindi Katz (1994; cf. Nast 1994) calls a ''space of betweenness'' which, although negotiated and changeable, recognizes embodied relations of power between adults and children, researcher and researched, advocate and inductee. This recognition foregrounds our inability to know children fully or to speak for them authoritatively, it understands that we are bodily bigger and often quite intimidating, but it does not foreclose our ability to uncover aspects of their experiences and politics. It is a recognition that enables the exploration of how identities are constructed and constituted in and through space without the baggage of either an autonomous subject or one who is theorized as simply a node in a matrix of power (Nelson 1999, 349). This is not to suggest that expressions of agency can be disentangled in power/discourse relations but to recognize that those relations are situated and mutable. It is a recognition that is presupposed by fieldwork (the lived experiences of the researcher at home and at work) that demands the theorization of children as concrete beings situated in material contexts rather than as uncritically and endlessly fragmented becomings. Young people, and those who study them, are situated subjects. The research process must comprise mutual recognition. The care with which researchers interact with young participants is not ''value added'' or something secondary but is rather a merging of sorts.

The central questions for understanding young people and how they apprehend the world are not disembodied or disembedded respectively from bodies and places. These questions require that we understand the ''maturation process'' from the start as a task that can only be accomplished collectively and contextually. The progress that a young person makes to lead a healthy life is read off transformations in the structure of a system of interactions and not off changes in the organization of individual drive potential, as Freud suggests. According to

Donald Winnicott's musing about play and reality, these transitions are characterized by a certain mobility and plasticity that enables and allows resistance.

Children embedded in object relations

"There is no such thing as a baby," writes Winnicott (1975, 99) in an essay on the mind and its relationship to psychosomatic illnesses. By this he meant that our understanding of infants does not exist apart from the environment with which they and we relate. He goes on to say "I was alarmed to hear myself utter these words and tried to justify myself by pointing out that if you show me a baby you certainly also show me someone caring for a baby, or at least a pram with someone's eyes and ears glued to it." Although clearly writing from a Western privileged perspective, Winnicott's point is that a child's bodily experiences – oral, anal, tactile, auditory – cannot be separated from and are always shaped by and given meaning through and within the child's object relations. The physical environment perceived by the child to exist simultaneously as part of and as independent of the self is understood in conjunction with a world of social and cultural practices, and the symbolic–imaginary, which may be entered into and manipulated to formulate unique relationships. Winnicott conceived of these as object relationships by which he meant both the conscious and unconscious relations that people have with places, environments and other people.[4]

Post-enlightenment objects

Recent feminist and post-structural accounts of "personal geographies" and constructions of self are profoundly shaped by the work of object-relations theorists (cf. Aitken 1994; Sibley 1995a, 1995b; Kirby 1996; Robins 1996; Aitken and Herman 1997; Bondi 1999). Developed as an attempt to revise regressive and individualistic aspects of Freudian psychoanalytic theory, object relations were originally conceived by Melanie Klein and Winnicott to render the self more social and playful. Their depiction of object relations is richer than those developed by Freud because other subjects are not converted into objects (e.g. mothers) and the recalcitrant material of narcissistic drives. Some feminists find object-relations theory less androcentric because it de-emphasizes the role of the phallus in child development and it downplays the role of the unconscious in the development of sexual identity (Flax 1990; Bondi 1996). Jane Flax (1990, 110) notes that much of what constitutes object-relations theory is more compatible with feminist post-structuralist views of the body than Freudian or Lacanian analysis because it does not require a fixed or essentialist view of "human nature." Going beyond the notion that Freudian id-instincts are problematic for the adaptation to some form of reality, object-relations analysts focus on the experiencing person, the self or ego, which becomes identifiable in the child as she or he receives the emotional effect of the id-demands. That effect is managed and manifested in the relationships that the child has with objects, loosely defined so as to include both people and places. The wrenching processes of self-identification that Lacan portrays as intrinsic aspects of desire – illusion, alienation and self-estrangement – are treated by object-relations analysts as expressions of self that can be inspected and transformed. In her concern with the development of the social self, for example, Klein suggests that infantile desires – many of which are violent, sadistic and paranoid – are associated with discomfort and so, countering Freud (and Lacan), she notes that mothers are not responsible for infantile fantasies or the emotions of their children. Rather, the infant's earliest experience of social relationships is when a caregiver provides comfort against hunger, cold and so forth. Moreover, any pre-oedipal one-ness with the mother is lost when the child develops a sense of borders and self-hood, and a sense of the social (Sibley 1995a, 6).

Winnicott, whose psychiatric practice was supervised by Klein for three years, argues against her focus on the child–mother dyad and other standard oppositions and antimonies such as subject–object, mind–body, nature–culture, man–woman, reason–emotion and so forth. Rather, his reflection is similar to that of Lacan in that he is interested in the relationships between symbolization and culture. Winnicott's perspective differs from that of Lacanian symbolism, however, in at least one important way: whereas Lacan sees language as a relatively immutable structure within which the politics of identity are played out, for Winnicott all forms of symbolization are fluid and flexible, allowing for multiple and divergent meanings and therefore subjectivities. This is not to say that Winnicott side-steps issues related to concrete relations between mothers and their children, but he makes troubling assumptions concerning how subjectivities develop around parent–infant relations (cf. Aitken 2000a). In particular, Winnicott stresses the centrality of mother–child relations in his rather annoying concept of the "good-enough mother." By suggesting that mothers can be "good-enough," Winnicott (1971, 11–13) refers to the "illusion" of a possible reality created through the mother. As I understand his use of this concept, Winnicott is suggesting that by being good enough, the mother stands in for what the child creates in her imagination (hence his use of the term "illusion"). The good-enough mother creates illusions that are also transitional objects that enable the child to play with, and reconfigure, her experiences without threat or challenge. Winnicott (1971, 12) underscores that the transitional object is safe and neutral because no-one asks the question: "Did you conceive of this or was it presented to you from without?" I'll say more about this quite startling point in a moment but suffice to say here that it certainly draws us away from the patriarchal strictures of Freudian psychology. Winnicott then goes on to emphasize that once the illusion is in place (e.g. the breast as a transitional object), the main task of the mother is disillusionment (e.g. weaning). Although some feminist writers influenced by object-relations theory may note the essentialism embedded in this problem and its example, they usually side-step the issue and move on to what they consider to be Winnicott's post-Freudian contributions (cf. Henriques *et al.* 1984; Flax 1993). I argue that the way Winnicott highlights a precise structural dialectic (illusion then disillusion) as good enough for certain developmental outcomes in children is problematic. In particular, it does not help mothers to distance themselves from a set of binding and perhaps stultifying moral obligations around the development of their children. It is worrisome to me that object-relations theorists do not engage the morality of these obligations or, in a larger sense, how certain myths of parenting and child development are established and sustained.

Being and becoming

With this caveat in mind, and there will be more, an important point of Winnicott's formulation of object-relations theory to what I am concerned about in the chapters that follow is that the nature of "childhood" necessarily changes as the objects around the child – social relations and family structures – change. The child not only "becomes" through the influence of cultural, social and political environments but she also brings something of herself "into creative living and into the whole [of] cultural life" (Winnicott 1971, 102). Common to both Klein's and Winnicott's work is the idea of a developing child through simultaneous inward and outward representations, and an emerging sense of borders and self, both real and imagined, as social and cultural constructions.

As noted in the example given above, for Winnicott, the whole social field is held together in a transitional space, relating subjects to each other and to reality while at the same time maintaining a certain amount of slack and flexibility. Transitional spaces are for "play" and reconfiguration, belonging neither to the subject nor (at least to some extent) to reality. In a paper that rehearses much of what I say here, Tom Herman and I (Aitken and Herman 1997) note that Winnicott's notion of transitional space messes up and makes fuzzy any attempt to establish the Cartesian mind/body dualism. Transitional space "is not inner psychic reality. It is outside the individual, but it is not the external world. . . . Into this play area the child gathers objects or phenomena from external reality and uses these in the services of . . . inner or personal reality" (Winnicott 1971, 51). Existing as a third type of reality that both separates and unites internal and external existence, transitional space represents a "neutral area of experience which will not be challenged," allowing for a flexible manipulation of meanings and relationships. Objects, places, cultural practices and self-images may become elements of transitional space and may be altered as an individual adjusts and updates knowledge throughout a lifetime. The empirical phenomenon that Winnicott has in mind here consists in the strong tendency of children from a few months in age to form highly affective, charged relationships with objects in their physical environments. He describes certain objects (bunnies, security blankets) as part of the transitional space because they are the first area of experience which is neither self nor mother. Hart (1984, 102) points out that because these objects help reduce anxiety, they are of great importance to the development of early geographies in that they serve as an aid to the child's exploration away from the nest of the crib and the arms of the mother. Treated as an exclusive possession, sometimes these objects are tenderly loved while sometimes they are passionately abused. The key to understanding the function of this transitional space is the fact that the child's partners to interaction also situate the objects in a domain of reality and so the question of illusion becomes testable. To use one of Winnicott's paradoxes, the infant must "destroy" the object in order to enter into a relationship with it. This is the process whereby the child removes objects from self-centered knowledge through their own agency: the object is destroyed and thus placed outside the arena of subjectivity. The survival of the object enables the child to perceive that he or she is not omnipotent and that the object has an existence "outside." In this way, a world of shared and meaningful reality is created. This can only occur once the object has successfully "survived" the child's destructive fantasies. The survival of the object, whether a person, a place or a favorite bunny, means that it can be safely hated, repudiated, and rebelled against, all of which strengthens the child's love for and dependence upon the object. Conversely, Winnicott implies that unless we tolerate the ruthless side of our character, it is impossible to have the full experience of the survival of the object. There is nothing in this to suggest a wildness predisposed to nature. Trust is the confidence gained by the object's survival of the child's destructiveness (Goldman 1993, xxii), but this process is also about testing borders and bodies.

Bordering the self

Julia Kristeva uses object-relations theory to answer questions about borders by focusing on the confusion and anxiety for a subject attempting to apprehend autonomy when faced with corporeal conduits between the internal and the external (e.g. nose, mouth, anus, penis, urethra, vagina). She argues that aversion to bodily excretions is a social construct that is metaphorically and linguistically linked to symbolic aversions to the "other" that are embodied in racism and sexism. Kristeva (1982, 32) uses Winnicott's notion of the "spaces of play" (transitional spaces) as provisional boundaries to help understand relations with the "other." In her rendering, transitional spaces are partial, containing semi-objects that are not quite real and not quite part of the subject. Kristeva's writing on "abjection" (hopeless anxiety because of something we do not like but cannot get rid of) supplements Winnicott's ideas with an emphasis on textuality and the role of language in under-girding personalities and the placing of the "other." David Sibley (1995a, 8) draws on Klein and Kristeva to suggest that the way some people construct themselves against a "generalized other" is abject in the sense that there is an urge to make a separation but this creates anxiety because such separations can never be achieved. Sibley's larger project suggests that abjection helps us to understand the geographies of exclusion that surround sexist and racist stereotyping. This relates closely to Nast's (2000) work although her project more forcefully articulates the relations between interiorized psyche representations and larger external political practices. Winnicott's perspective is perhaps more liberating because from the outset, and prior to any formulation of an oedipal or mirror crisis, is the recognition that the self is enmeshed in relations with others. Importantly, what holds together these relations is the power of illusion (belief, historical pattern, geographical contingency) to bind individuals through what Glass (1995, 174) calls "a set of shared assumptions which actively protects the self." It is a humane argument, Glass goes on to argue, that unshackles itself from the phallocentric center of classical humanism and psychoanalysis.

Unlike Freudian and Lacanian views, theories of object relations presuppose an active and positive experience with objects and places rather than a world in which there are not only significant tensions between self and other, but also gaps that can never be bridged. The transitional space is the psychic environment within which some interpretations are formed and used, where others are generalized or not. Winnicott emphasized the notion of this as a "holding" or "facilitating" environment, but what is really crucial is the understanding that transitional space is precisely the interplay between the internal and the external. Interpretations emerge from neither the child nor the object, but from the movement of child to object and object to child. The idea of a transitional space, then, relates fully to Gloria Anzaldúa's (1987) "body as activity" in resistance, survival and geographical transformations (cf. Zita 1998, 168). The "reality testing" function of transitional objects, Winnicott writes (1971, 13–14), persists in shared illusions or what he calls more generally "culture." Given that transitional spaces are spaces of "play," then it is out of these spaces that culture and social transformations arise from the movement of children to culture and culture to children.

Culture and symbolism

Culture and symbolic-imaginaries are not immutable structures that define children, and children are in the process of creating future cultures and symbolisms. According to this theory, culture is not conceptualized as Freud's or Lacan's external and coercive "law of the father" which forces the child's separation from the mother. Rather, the child is able to bring something of the inner self to the traditions and practices of society in order to be able to make use of them. The agency of the child shapes his or her cultural practice. Also, the child's ability to choose and utilize transitional objects points to the beginnings of the process of symbolization. This differs from Lacan who positions symbolization at the point of language formation. For Winnicott (1971, 102), the capacity to play and the process of symbolization expand "into creative living and into the whole cultural life of man [*sic*]."

Culture, like play, is not only something that the child can "make use of," but it is also a tradition to which she can bring something of her inner self. It is not difficult to take this reasoning a little further and infer the creation by individuals of places that symbolize important aspects of self. There is an important geography here: space becomes place for the child with all its attendant symbolism and politics. The dialectic nature of this process moves us beyond binaries, and establishes an important connection between the ideas of Winnicott, Lefebvre and Nast. It also provides a conduit to the next chapter, which moves beyond the psyche to focus on the material conditions of children's experiences. Spaces and places are important not only because they embed and contextualize children, but because they enable an important form of corporeality through which sex, race and culture are experienced rather than imposed.

Geography, perhaps more than any other discipline, can make important contributions to our understanding of the situatedness of children and their "doing" of identity. In doing identity, children break into geography, culture, ethnicity, morality and a host of other pretensions. One possible way to avoid the traps of the Cartesian subject that I have sketched out in this chapter is to examine the place of children in a given society as a whole, without exempting them from the larger multi-faceted transformations that societies undergo. The embeddedness of children that needs broaching now relates to their relations – and the relations of childhood – to larger social transformations. In the next chapter, I try to think through how childhood is materially spatialized and historized as it creates and recreates a constitutive part of children's identities. Indeed, I argue that it is imperative to understand how the ways that we conceptualize childhood stand in for those transformations and what is expected of children as part of the transitional space of culture. As Stephens (1995, 18) points out, a crucial part of describing the distinctive shape of contemporary global culture in different world regions and social contexts is exploration of the pivotal figure of the child and the relation of this ideal construct to the diverse lives of children. What are the implications for society as a whole, she asks, if there are no longer spaces for children that are not conceived as at least partially autonomous from the market and capital-driven forces? And what happens to the bodies and the minds of children in the process? The arguments in this chapter for a fuller problematizing of children's racial and sexual identities join with my hints at the end of the last chapter that their disembodiment serves global capitalism. The point that I pick up in the next chapter is that children's minds and bodies are laid bare as flexible "natural" resources to be reshaped and commodified, sometimes without resistance, as larger social, political and economic transformations dictate.

Notes

1 More concerned with unconsciousness than physiographic function and embodiment, Freud's smallest possible entity was the birthplace, or progenitor, of memory and consciousness (Steedman 1995, 77). He was influenced by the ideas from functionalist physiography in the sense that he believed entities were composed of smaller parts, but he drew a line between its reductionism and the notion of the foreskin as a separate structure as described in Chapter 3. None the less, as a student of his own heritage, Freud understood that ritual Jewish circumcision was meant to be a traumatic event precisely because its purpose was to inhibit male sexuality. He argued that circumcision was a modern symbolic substitute of castration, "a punishment which the primeval father dealt his sons long ago out of the fullness of his power" (Freud 1939, *Moses and Monotheism*, quoted in Gollaher 2000, 67). By the mid-1890s, Freud was an advocate of vigorous sexual expression and he was convinced that physical and emotional repression of sexual arousal and release could provoke neurosis.

2 Although not as commonly discussed as masturbation, when made part of the public record, adult sexual relations with children were also documented and examined with care. In Victorian London there are numerous court records on sodomy cases and cases against pederasts. Many of these ledgers provide explicit details on places for, and changes in, the intimate sexual relations between teenage boys and men (Trumbach 1999). The larger project was, of course, the control of homosexual men and the confinement of their activities to only specific parts of the city (public toilets, parks, etc.).

3 Other commentators of the time, pediatricians prominently among them, warned parents that masturbation was learned in infancy and that the foreskin was chiefly to blame. According to J.B. Webster, for example, boys develop the habit when they are about a year old, "due in the first place to the condition of the prepuce" (Gollaher 2000, 104). As alluded to in Chapter 3, it was with these kinds of discourses about controlling the seemingly improper aspects of a boy's sexuality that circumcision gained popularity.

4 "Object relations" is an awkward term adopted by psychoanalysts from the ground-breaking work of Melanie Klein. Her depiction of object relations is richer than those developed by Freud because other subjects are not converted into objects

(e.g. mothers) and the recalcitrant material of narcissistic drives. Important for what I argue here is Klein's concern with the development of the social self. Object-relations theory developed from the Kleinian school to encompass several perspectives. In the United States, from 1920 to 1950, Harry Sullivan used it to develop his interpersonal school of psychoanalytic thought, as did Ian Suttie and Henry Guntrip. Esther Menaker (1995) argues that the term ''object relations'' derives from grammatical usage, where people are simultaneously the subject and object of verbs (quoted in Buchholz 1997, 314).

5

MATERIAL TRANSFORMATIONS
Local children in global places

There is a rapid proliferation of literature theorizing the processes that underlie contemporary global transformations. Whether the moment is framed as globalization, late capitalism, post-industrial society, or a global extension of flexible accumulation, analyses of childhood and the experiences of children are largely missing from current theorizing. A focus on children and social reproduction is important, argues Stephens (1995, 8), insofar as they "break the frames of dominant models of transformations in the world system." The minds and bodies of children, and the ideal construct of the child, are a crucial part of describing the distinctive shape of larger material transformations, if for no other reason than that they embody the desires and dreams of adults. Our judgements as to what matters in being an adult explain in large part why we have particular conceptions of childhood, but the reverse also holds: particular conceptions of childhood reveal a large part of the social imaginary of adults. The purpose of this chapter is to discuss the creation of childhood and adolescence as material and commodified parts of the modern period of Western industrial capitalism and then elaborate on how these notions are exported globally. In so doing, I take issue with some contemporary commentators' insistence that "childhood" and "youth" disappear as categories of experience towards the end of the twentieth century. Rather, I argue that indeterminacy in what constitutes contemporary childhood and adolescence parallels processes of globalization. At its most crass, this indeterminacy marks children as flexible – perhaps the most flexible – consumers and producers of capital. At its most buoyant, it enables enlarged spheres of identity choice and the possibility of an ecumenical political will for young people.

I assume that the so-called nature of children changes as the objects around young people – social relations and practices, and family and community structures – change. Children not only "become" through the influences of these changing objects, but they also bring something of themselves into cultural life as they actively participate in the day-to-day workings of places. The notion of a fluid transitional space is the springboard from which the current chapter finds an initial foothold to propel it into a discussion of the complex beliefs, feelings and sentiments that surround children, childhood and the self. Pivotal questions revolve around what happens when children lose access to transitional spaces with which they can tackle, embrace or destroy ideas about self and others. Winnicott notes that transitional spaces reside somewhere between our interior selves and the exterior world, and it may be argued that as a space of justice, this third space cannot be challenged. I argue that it is indeed under assault by commodification and mechanistic ways of knowing. The questions that permeate discussion here revolve around whether this space atrophied with the kind of instrumental global reasoning and logic that has come to predominate Western society through the last century.

At least in some ways, children are now closely identified with adult selfhood and I argue that this conception creates some problems for people who are "in a state of childhood" in two ways. Following the overarching theme of what is happening to children's minds and bodies, I argue first that there are important reasons why we need to understand the ways childhood and adolescence are represented. This comes in part from the transformation of adult attitudes from indifference to a recognition of children as different in modern times. My argument is furthered by problematic new forms of indifference in post-modern notions that see contemporary childhood disappearing in an homogenization of symbols that fail to prescribe what it is meant to be a child, adolescent or adult. Second, I outline an argument for the positioning of childhood as a global political lightning rod that attracts and absorbs powerful, typically negative feelings and reactions, thereby diverting interest from other issues. The lightning rod varies in how it depicts children but it highlights the importance of understanding the work and experiences of young people as part of the transformation of social reproduction.

Transforming childhood

Ariès is doing the rarest kind of history, a history of the present, aiming at de-naturing "the child," exposing our own constructing apparatus, freeing us, at least a little, from the tyranny of our eccentric seeing.

(Kincaid 1992, 62)

It may be surprising to some that it has taken me so long to get to the work of Philippe Ariès, arguably the most important progenitor of modern historical constructions of childhood. I have held Ariès' work in reserve until now because I want to use it in a very specific way. I want to dispel the eccentricities of positivist debates that construct childhood as some kind of a natural stage and in their place outline material transformations that have bearing on the construction of child-hood as a moral imperative. As James Kincaid (1992, 62) notes, Ariès drives ''a sharp wedge between the child and nature, shown as the contingent, deter-mined nature of this phenomenon, the child.''

Most scholars agree that Ariès' work established that childhood was not always as we know it in Western society, but emerged as part of the early modern period of industrialization. If this is taken to mean that there simply is no child in the past, then the familiar developmental debates of the previous chapters hold sway. Rather, in *Centuries of Childhood*, his seminal work on the topic, Ariès argued that prior to industrialization, children ''did not count'' (1962, 128) because the prevailing high levels of mortality did not inspire affection or interest in a being whose hold on life was quite tenuous. This is an argument that speaks to material transformations and emerging moralities. With industrial capitalism and the material transformation of public and private realms, childhood became recognized as a separate sphere of experience. And, as I alluded to in the last chapter in reference to Lefebvre's (1991, 49) work, the social body and the body of the child is first recognizable with the separation of productive and repro-ductive actitivies.

Centuries of Childhood is generally accepted as a rich and definitive work on the creation of modern childhood and it initiated the idea that childhood is a problem-atic social construction. As an historical benchmark, it provides a useful starting point for the understanding of contemporary geographies of childhood because it marked the time when indifference to children was re-*placed*.

Indifference towards children

Ariès is not without his critics. A number of historians raise issues about his methods, and there are some problems with his assumptions (cf. Wilson 1980; Pollock 1983). Early on, for example, Lawrence Stone (1974, 28) argued that his methods were so flawed and his evidence so overwhelmingly inadequate that inferences should be viewed with suspicion. But *Centuries of Childhood* is noteworthy today because Ariès used materials that were not at the time con-sidered appropriate for the historical rendering of subjects. He employed sources such as paintings, costumes, architecture and literature, and investigated games and attitudes towards sexuality with considerable interpretative skill. *Centuries of Childhood* is now accepted for the most part as an historiography of family life that renders childhood accessible in qualitative, human terms where inter-pretation of art and games is equally as valid as numbers and authoritative texts.

From a different tact, Ariès' generalizing about parent–child relations was severely criticized by Linda Pollock (1983) who asserted – through sociobiological theory, anthropology and the study of primates – that parents are more often than not loving and affectionate towards their children, and always start off with infants they wish to raise to independent adulthood. Although weighed down by her own generalizations, Pollock none the less garnered significant support. In his own defense, Ariès points out that "the idea of childhood is not to be confused with affection for children: it corresponds to an awareness of the particular nature of childhood, that particular nature which distinguishes the child from the adult, even the young adult" (1962, 128). Comments such as this initiated the idea that childhood is a problematic social construction and, of note for the tenor of the discussion I make here, the "nature" that Ariès alludes to is about *links* and *inter-dependencies*. For example, Ariès (1962, 369–7) interprets relations between masters/adults and apprentices/children in terms of "existential bonds" and describes this kind of interdependence as a valuable system of human community. In many ways, then, *Centuries of Childhood* is about the work and play of parenting and how children are constituted as parts of that work and play.

Ariès' work is important in that it positions historically and materially the transitional spaces of childhood. But the work's tendency to generalize broadly from examples drawn exclusively from pre-industrial France and, to a lesser extent, England is none the less problematic. There is a geographic and chrono-logical vagueness in the work so that given points in time (e.g. the beginning of the seventeenth century) or specific places (e.g. urban, rural) are articulated with little precision. What is missing is a concerted effort to engage local contexts of change in specific and concrete ways. To a large extent, this is the agenda of current empirical studies of children's geographies but before I get to that discus-sion I want to situate the theoretical importance of Ariès' work on childhood and link it, specifically, to material transformations.

Ariès (1962, 411) argues that pre-industrial "collective life carried along in a single torrent all ages and classes, leaving nobody any time for solitude and privacy." Community life was public life, and the function of families was to ensure the transmission of life, property and names: "In these crowded, collective existences," Ariès writes, "there was no room for a private sector." Nothing was separated out from the vantage of individual, family or community, and children, from the time they were weaned, were considered quintessential parts of public space. They participated in adult society because no special provisions were made for their welfare. As partial evidence to support this conjecture, Ariès (1962, 43–6) notes that prior to the mid-eighteenth century portraits in upper-class society depict children as little adults not only by their clothing but also by their physiognomy. This is not to say that children had the rights and power of some adults, or that particular attitudes pervaded adult–child relations, but that indifference rather than difference marked those relations.

Children as different

According to Ariès the first concept of children as different arose in the early modern period and, importantly, it relates to *mignontage*. As I noted in Chapter 3, this concept literally means to ''coddle'' or bind but its meaning is transformed when children's antics were seen by women (''mothers, nurses and cradle-rockers'') with delight, almost as a source of recreation and relaxation. First women, and then some men, recognized the pleasure they got from watching children and indulged in coddling and pampering them to accentuate the levity. Given the timing of these transformations, it is important that the concept of coddling is tied to Goethe's conceptualization of the child-figure Mignon. As I noted in Chapter 3, Steedman's (1995) argument is that it was from this period that adult notions of child-like interiority (the inner child) arose but it is also important to remember Goethe's child-like figure as one that is dislocated, misplaced and eroticized.

By the end of the sixteenth century, Ariès avers that many male commentators found insufferable the attention paid to children and called for instilling discipline and manners in children rather than indulging in distraction and fussing over them. Moralists and pedagogues of the time (quoted in Ariès 1962, 130–1) did not regard children as charming toys but as fragile creatures/animals of God who needed both safeguarding and reform. Seventeenth-century attitudes stemming from Puritanism constituted the child as anarchistic and fundamentally evil, requiring parentally imposed strictures and tutelage that were God-given and absolute. In the Judeo-Christian tradition of contriving a garden out of the wilderness, childhood was seen as a time for laying waste to natural urges and the cultivation, usually through the father, of moral values.[1] An attack of insipid naturalism and the spoiling of children is noted in some of the diatribes against coddling: ''I cannot abide the passion for caressing new-born babies, which have neither mental activities nor recognizable bodily shape by which to make themselves lovable . . . for our amusement, like monkeys,'' wrote one commentator (Montaigne, quoted in Ariès 1962, 130) while another avers that ''it is as if the poor children had been made only to amuse the adults, like little dogs or little monkeys'' (Fleury, quoted in Ariès 1962, 131). These ''moralists and pedagogues'' did not regard children as charming toys but as fragile creatures of God who needed both safeguarding and reform (Ariès 1962, 133).

Adrian Wilson (1980, 134–5) points out that it does not matter whether childhood is regarded as a state of weakness and imperfection, or one of innocence and grace, what matters is the development of ''reason,'' and this was based on a specific understanding of children's natures that became associated with child-centered pedagogy. Children were different from adults, their constitution was separate and malleable: boys were instructed in rationality and logic and girls in etiquette and moral duty. Reason dictated further the disciplined place of men and women in the public and private spheres respectively. An important geography of the early modern period, then, relates to the material evolution of separate public and private spheres. Whereas *mignontage* derived from the intimacy of the family and the community, the new moralist pedagogies of dislocation were derived from the public realm (particularly the church and psychology) and "passed into the family" (Ariès 1962, 132).

> The reformers, these moralists . . . fought passionately against the anarchy of medieval society . . . and the transformation of the free school into the strictly disciplined college. This literature, this propaganda, taught parents that they were spiritual guardians, that they were responsible before God for the souls and indeed the bodies too, of their children. Henceforth it was recognized that the child was not ready for life, and that he had to be subjected to a special treatment, a sort of quarantine, before he was allowed to join the adults. This new concern about education would gradually install itself in the heart of society and transform it from top to bottom.
>
> (Ariès 1962, 412)

The geographic question that Ariès ignores turns on how this transformation occurred. The reconstitution of space over a long period, from roughly the sixteenth to the nineteenth century, into separate public and private realms contributes to Ariès' thesis on the way that childhood emerged. In short, the wrenching of public workplaces from private domestic spheres propelled change in major social processes including the emergence of a spatially separate family identity. In the nineteenth century, after some struggle through various factory and labor laws in Europe and the United States, the place of the child was set squarely in the private sphere along with women and, at least initially, education and health issues.[2] The modern family evolved as a private retreat from sociability and so the "discovery" of childhood prescribed important changes in family and social structures. New forms of intimacy between parents and children were initiated with the creation of working-class and middle-class home life in Western society as increasingly private and separate.

By the end of the nineteenth century the attitude of indifference towards children had undergone a radical transformation. The association of children and parents became increasingly stretched and, in places, broken. With the growing belief that the future rested on children's shoulders, certain aspects of societal change were deemed too important to be the sole responsibility of parents. Education and health were institutionalized as part of the state apparatus with the admonition that family life must be focused on moral issues and nurturing.

Ariès (1962, 412) notes that education was the great event that heralded a move towards the recognition of childhood. Neil Postman (1982) extends Ariès' concern by asserting that education separates out children in order to prepare and socialize them for the adult world and in this way it becomes a marker for the transition to adulthood. The amount of time children and youths spent in educational institutions increased dramatically through the twentieth century and, as I have noted, there is significant discussion in the literature about the implications of this separate sphere (Wolfe and Rivlin 1987; Thorne 1994). Archard (1993, 29) reiterates Ariès' observation that the most important feature of modern Western society is that it conceives of childhood as meriting separation from the world of adults but he notes further that this is also a separation from rights in justice and law: "Children neither work nor play alongside adults; they do not participate in the adult world of law and politics." To understand the context of this separation, I want to consider the disciplining of play in the last century because it relates most cogently to my concerns about the erosion of transitional space.

Elizabeth Gagen's (2000a, 2000b) study of the Playground Association of America (PAA) is particularly insightful in this regard because of its focus not only on children's play and social transformation but also on the spectacle of young disciplined bodies. This is important for understanding processes of globalization because the export of transformed notions of childhood, socialization and education is inextricably linked to the export of modern constructions of gender, individuality, and the family (Stephens 1995, 16).

The surveillance of play and the disciplining of bodies

By the beginning of the twentieth century, children's play is not only constituted as important (time spent in factories deprives children of appropriate education and of other childhood experiences) but it is separated out from the sites of adult productive activities. Supervised playgrounds were instituted across Europe and the United States as a means of drawing children off urban streets and into corrective environments. Gagen (2000a, 2000b) points out that the theoretical construction of the child that reformers drew from, and indeed that made the logic of playground reform cohere, made it necessary to display children in public because they recognized children as a means of larger social transformation. She argues that reformers believed in the plasticity of children over adults, and that the child's body was constructed as a conduit through which identity surfaced. Public education ensured the disciplining of the mind, and disciplined bodies in playgrounds as part of public spectacle further induced the transformation of internal identities. That this material transformation is about spectacle is important for Gagen's argument and resonates with Robins' (1996) discussion of the changing moral and visual spaces of this century.

125

Gagen (2000a, 601) points out that "the problem of the children on the street" was first brought to the attention of middle-class Americans with the publication of Jacob Riis' *How the Other Half Lives* in 1890. This book was famously suffused with visual imagery of grimy alleys, dank buildings and, importantly, the gaunt and seemingly depraved looks of ragged young people. The popularity of Riis' book prompted a lecture tour that added live lantern performances to the book's spectacle of visual images. This was a clear exoticization of the abject, wild, child-like other. For middle-class viewers the lantern slides and narrative provided a prototype of virtual tourism where the slum could be viewed safely and without accompanying dirt, odious smells and human vulgarity. Perhaps most importantly for what consequently transpired at the beginning of the twentieth century, paired lantern slides enabled a simultaneous inspection of children before and after salvation. Riis often presented juxtaposed pictures of children before they entered care institutions looking suitably abject with pictures of the same child after rehabilitation (Stange 1989, quoted in Gagen 2000a). A solution to what might be done with Mignon is now offered in the form of saving children from delinquency by removing them from urban degradation. Paralleling Riis' campaign in the United States, Margaret McMillan was able to put childhood on the political agenda of the British Labor Party. Like Riis, McMillan described striking vignettes that allowed her readers to "see" the need for cleaning up a notoriously poor and "derelict" part of south-east London. Her political agenda was to justify the operations of an open-air school and medical center in the area. In one particularly poignant pamphlet entitled "Marigold – An English Mignon," McMillan highlighted the physical plight of the 7-year-old child of a coster (Steedman 1995, 2). The removal of urban degradation and dereliction was to get children seen, out in the open.

The spectacle of depraved Mignonesque children was now at the forefront of the popular middle-class imagination; transformation of a particularly spatial and material form was possible. Riis and McMillan exposed the danger of leaving hidden the spaces thought to be responsible for altering the character of the population. Importantly, this visual spectacle was combined with scientific surveys and data on infant mortality to provide an impetus for moral/spatial change. For example, following the example of the sociology department of the University of Chicago, many universities established "settlement houses" in twentieth-century slums from which "dozens of researchers . . . scoured the area for information, much of it highlighting the mostly awful aspects of local life" (Jablonsky 1993, xiii). As Nast (2000) points out in her careful documentation of the spatial practices of the Univeristy of Chicago, larger issues encompassed the control of black bodies and the colonization of their spaces. These spatial practices ensured a bounded and sanitized space from neighboring communities as they experienced significant industrial-led in-migration of southern African-

Americans but, as I noted in the last chapter, the spatial practices were also about infantilization.

Surveys, observation and other kinds of scientific data collection fueled the engine of change but it was the accompanying visual imagery that sparked the impetus to save children from immigrant communities that were increasingly seen as depraved and sordid. It is no coincidence, of course, that these communities were also becoming less white. It made sense, argues Gagen (2000a, 602), to bring children out of these squalid communities "and playgrounds provided that perfect opportunity to lure children into the open." The important point about this change is that it presupposes a focus on children's bodies that is traceable back to Goethe.

The positioning of the children of the urban poor in this way was important for two reasons. First, seemingly unruly children were placed under a controlling public eye and, second, their disciplined bodies were put on display as show pieces to attest to the social changes underway. The notion of combining discipline and spectacle, argues Gagen (2000a), takes the Foucauldian project beyond that of discipline through surveillance to a more proactive form of social transformation that targets material bodies and takes control of social reproduction. With playgrounds, spectacle, surveillance, discipline and the control of social reproduction are interwoven in the same project. The importance of spectacle is validated with Riis' appointment as honorary Vice-President of the PAA when it was founded in 1906.

The grounding theoretical basis of the association derived from neo-Lamarckianism and recapitulation. Gagen notes that although progressive thinkers like John Dewey, James Baldwin and William James influenced the PAA, it was Granville Stanley Hall's (1904) theories of child development that focused the movement on crucial aspects of environmental control. The influence of recapitulationism in Hall's work was to establish a form of environmental determinism that influences playground design to this day. The founders of the PAA believed that children must successfully reenact each evolutionary stage to develop correctly, and Hall's rhetoric is used to establish the need for playgrounds:

> There is practically definite growth and development of the child, physically, mentally, socially, morally, along lines more or less parallel with the development of the race, and that one familiar with these facts of child development can gauge with considerable accuracy the interests, aptitudes and needs of children of a given age.
>
> (Johnson, quoted in Gagen 2000a, 605)

If the instincts of children at particular ages were gauged accurately then they

Figure 5.1 Contemporary notions of surveillance (eyes-on-the-playground) and a particularly poignant patriarchal nursery rhyme are exemplified in this playground in Singapore

could be controlled through play rather than be allowed to run wild. If children were allowed to develop these instincts on their own without supervision or the correct environmental provision, then the benefits of the instincts would be lost to anti-social behavior. So, for example, the instinct of adolescent boys to club together in street gangs (tribes) could be redirected towards team sports on the playground. Part of the control of social reproduction is constituted in the supervision of play and the establishment of recreational programs for boys and girls. The public space of the playground displayed ideal male identities ("play as the moral equivalent of war," George Johnson 1912, cited in Gagen 2000a) and female identities (the display of dance, drills, song and crafts).

The important features of Western society during the first half of the twentieth century, then, include its conception of children as meriting separation from the world of adults, education as an event that distinguished childhood, and the public display of particular kinds of behavior in boys and girls.

The death of childhood

Playgrounds may be thought of as part of a material transformation that increased institutional control over children's lives in complex ways. Ironically, the myriad ways in which children's leisure became organized and contained in the global north after the Second World War parallels the extension of institutionalized educational control and changes towards requiring a longer period of formal qualifications in order to enter the workforce. As Kincaid (1992, 69) speculates, adolescence became "inconveniently lengthy." Postman (1982) argues that this extension of childhood visually, spatially and temporally produced a situation in which clear-cut cultural divisions between childhood, youth, and adulthood once more evaporated. He says that the decline of American childhood as a protected familial space began in the 1950s. In contrast, in pre-industrial Europe and continuing through modern Western society, having children often heralded full membership to the adult realm. In the United States and a large part of Western Europe, up to and extending well into the 1950s, marriage and children followed graduation from school. The popularity of June weddings is perhaps a remnant of this particular rite of passage. After the Second World War, the specificity of this rite eroded but early marriages (made possible by housing subsidies and rising standards of living) and restricted access to sexual relations still produced a situation in which (for men) entry into the workforce, marriage, and a full sexual life all occurred within a relatively short period. Ivar Frønes (1994, 152) argues that this indicated a "natural phase" for the passage to full adulthood which became less identifiable from the 1960s onwards in many Western contexts.

In the latter half of the twentieth century, then, the inculcation of family values in the home and community values in the school gave way to an uncontrolled invasion of children's minds by market-driven media images and globally circulating signs.[3] This invasion, argues Postman (1982) and others (Hengst 1987; Cunningham 1995; Jenks 1996) signals the disappearance of childhood as a distinctive stage because it heralds the loss of innocence, particularly sexual innocence. If there is no longer a convergence that produces identifiable distinctions between a prescribed experience of childhood, adolescence and adulthood then what moral implications arise from their disillusion? Academic and legislative concern is illustrated by a sustained emphasis over the last quarter century on children's formal rights (cf. Adams *et al.* 1971; Okin 1989; Archard 1993; King 1999), the weakening of parental or household authority, and the growth of an ideological predilection for individuation that speaks to material transformations:

> A room of one's own for a child is not only a private spatial sphere made possible by increased affluence: it is also a private symbolic sphere,

underlying the child's position as an individual and a personality. The individualization of childhood pushes the common categories of ''children'' and ''parents'' more into the background and stresses the intentions and personality of the individual child or parent.

(Frønes 1994, 153–4)

Frønes goes on to argue that the culture of the ''democratic'' modern family is characterized by negotiation and the homogenization of symbols through which decision-making and social control take place (cf. Elshtain 1990; Wood and Beck 1994). For Postman, the homogenization of symbols is primarily through the power and pervasiveness of contemporary media, but the destruction of the childhood realm is also traceable to changes in public education, methods of upbringing, and how families are formed. What is important here is, of course, a set of material transformations that relate to the ways reproduction is continually reconstituted by changes in larger structural forces.

Globalization and the reconstitution of reproduction

The account of the complex ways in which childhood has been transformed that is touched on above finds some current coherence in the notion of children as artifacts of contemporary globalization. Globalization is not a new phenomenon, although its pace and effects may be. Contemporary globalization began with the drive toward capital accumulation requiring an ongoing expansion of exploitable labor pools, the constant expansion of markets, and the quest for new resources (and places) to exploit. It included disinvestment in certain geographical locations, de-industrialization, the decline of fordist forms of manufacturing, and the rise of economies based on the service sector and high technology. The quest for new forms of investment led to particular forms of industrialization in previously non-industrialized countries and the concomitant devastation of subsistence economies. In the same sense that the processes of globalization are neither unidirectional nor uniform then positioning childhood is indeterminate because the local conditions of global children are so varied. All cultures to one degree or another give meaning to physical differences in age and sex, but what I am arguing here is that the particular form of the modern child is socially, historically and geographically specific. Ariès' empirical work was geographically limited but it spoke to a specific moment in the construction of modern childhood and suggested its emanation from Western Europe. This is not to make monolithic claims of terms like adulthood and childhood as they relate to different cultures, but the processes of globalization presuppose a world of Western assumptions. The nature of childhood, then, requires an elaboration of the nature of globalization.

130

The globalization of children is not natural, but contrived from processes of fluid capital accumulation. The early industrial era split productive activities from reproductive activities, but these two sets of processes require contiguity. The domestic sphere was the site of reproduction, nurturing and disciplining a new labor force that was kept healthy and appropriately educated by local institutions. Globalization changed all that. Capital now has fewer commitments to reproducing any particular labor force or conditions of production (Katz 1998). Geography texts are replete with stories of disinvestment and what that means for adult labor (Mitchell 1993; Natter and Jones 1993), but the effects on local children are not as yet fully articulated. The important point about contemporary globalization from the standpoint of children's geographies is that there is no longer a need for capital investments to be secured at particular locations.

Cindi Katz's (1991a, 1991b, 1993, 1998) project is to re-introduce social reproduction as an important, but as yet missing, aspect in the globalization debates. Production cannot exist without reproduction, which is not only about biological reproduction but is also about the daily health and welfare of people. In addition, social reproduction is about the differentiation and skill of the labor force. Labor's struggle through this century over wage, education and health advancement produced costs for capital that ultimately led to significant relocation and disinvestment. Children's geographies are integrally linked to social reproduction but the places and practices of children's everyday lives are rarely considered to be a dynamic context for understanding historical change, geographic variation, and social differentiation. For example, feminist geographers use the concept of social reproduction to analyze the day-to-day lives of women, and how they contribute to maintaining productive activities through non-waged labor in the domestic realm (Gregson and Lowe 1995; England 1996; Holloway 1998a). While focusing on issues such as child-care, child-rearing and the space of the home, these discussions speak only of the ways that social rules, ideals and practices are transmitted to children, but they say very little about the ways that they are received, internalized, resisted and mobilized. Young people provide a different, and in many ways more illuminating, window onto social reproduction. It is during childhood and adolescence that the principles of society are mapped onto the consciousness and unconsciousness of embodied subjects. It is also when some portion of social reproduction is contested and negotiated, and this resistance is most often embedded in local geographies.

Economic restructuring in New York and Sudan

In one of the first attempts by a geographer to understand the specific effects of globalization on social reproduction, Cindi Katz (1991a, 1991b, 1993, 1998)

Figure 5.2 Boys' agricultural work in rural Sudan
Source: Photograph by Cindi Katz used with permission

focuses on the similarities and differences between the lives of children in New York and a rural village in Sudan. Her aim is to provide insight into the notion of economic restructuring and global change by demonstrating some of its local consequences. In New York, she discusses school and peer groups as public and neighborhood based social practices used by young people to construct themselves and make community. Local disinvestment in New York and other urban areas in the United States has for years been a prime catalyst for destroying neighborhoods and fragmenting extended families. The authority of the family has been eroded by public and private institutions that remove children from "unfit" parents and attempt to teach proper parenting, or provide mentors and role models (e.g. "Big Brothers") where none is perceived to exist. Katz (1991b) points out that concern here is not for child welfare or there would also be institutional intervention in the family life of the middle and upper classes where there is equal evidence of neglect, abuse and dysfunction. Rather, concern is for control of the social reproductive system of the so-called lower classes and to tie them more closely to the prevailing social relations of production engendered by late capitalism. Katz reads traits such as the development of gang-communities, the security of teenage pregnancy, and the refusal of youths to take on waged employment as part of a larger pattern that threatens the regime of capital accumulation.

Figure 5.3 Girls' agricultural work in rural Sudan
Source: Photograph by Cindi Katz used with permission

She describes the ways that the New York school system historically excelled in allocating class membership and how school-leavers were equipped in the past to take on steady waged employment which was reasonably well-paying and community-based. Since the social and economic crises of the late 1960s, however, the public education system in New York has broken down due to disinvestment and the removal of flexible capital. Katz (1991b) points out that business and industry has attempted to coerce the education system into re-educating its recalcitrant youth. Business and industry complains through the popular press that the schools are not preparing children with the proper attitudes and work ethics (rather than vocational skills and knowledge) for the labor force. Companies and private foundations have established costly business–school partnership programs but the New York City Public Schools remain unable to provide economically viable workers.

Katz's (1991a, 1993) work on changes in social reproduction in rural Sudan focuses on a large-scale local irrigation and agricultural development project. This state-sponsored and internationally financed project transformed social relations of production and reproduction as well as the local political economy. Amongst other things, the building of a large dam and the implementation of new forms of agriculture were meant to increase tenants' productivity and thus reduce the amount of time that children labored in the fields and surrounding countryside so that they would spend more time in school. Ten years after the creation of the project, Katz (1993, 95) found that children's labor time and space had actually expanded due to the expansion of the cash economy, the degradation of the local environment and their mothers' changing work patterns. The agricultural project transformed 85,000 acres to cash crops requiring children and others to forage further afield for fuelwood, berries (primarily the work of girls) and grass to graze livestock (primarily the work of boys). In addition, children became involved with the cash economy selling water and fuelwood and working as paid field-hands. As a consequence, Katz (1991a) notes that the gains in school attendance a decade after the creation of the agricultural project were not nearly as large as expected. Of related importance is her finding that young men were increasingly likely to leave home in search of paid employment elsewhere. Male out-migration led to yet another increase in child and female labor in local agricultural production. One of the central arguments that arose from Katz's research in Sudan (and New York) was that children and adolescents were not learning the kinds of things that they were likely to need in adulthood to survive in a shifting and fluid global economy. Because of the changes in the Sudanese village, young people were working long hours in a variety of subsistence and cash-oriented activities but the fulcrum around which they worked was the village system and this was itself under erasure. The agricultural project was eating up so much land around the village that it was unlikely that the young people would have access to productive land when they came of age.

Amongst other things, Katz's (1991b, 1993) work shows that the sexual division of labor helps us understand better children's work activities as they relate to social reproduction in global south contexts. She is concerned that because of changing and unequal social and economic relations, certain girls will be deprived of the chance to develop any kind of spatial skills which might enable them to negotiate their lifecourse. It is clear that girls' access to space in rural Sudan becomes severely constricted after puberty when the space that females control is circumscribed by the home. At this time, and through the child-bearing years, they are removed from all productive activities in agriculture. Older boys, on the other hand, get to learn a wide range of farming practices and field techniques. Although this constraint on girls is the result of *purdah*, or assuming "the veil" in Islamic cultures, Katz (1993) points out that women in the post-industrial West are similarly constrained in their access to productive activity and the control of space. While *purdah* is recognized as an explicit means of exercising patriarchal control, equally powerful codes govern female behavior in the post-industrial West. Most conspicuous amongst these are differential wage-rates for women and female harassment in a male-dominated workplace, as well as an engendered fear of crime in women which, however real, also reinforces patriarchal authority. The result, asserts Katz (1993, 104), is that girls in urban America are restricted from exploring and engaging with their environment perhaps even more than girls in ostensibly more restrictive societies such as Islamic Sudan.

The irony of Katz's comparison is that the development of the agricultural development project in Sudan resulted in significant changes in the lives of the girls who lived there. When older boys left the village to seek work, many of the adolescent women who had withdrawn from agricultural activity at puberty and had not learned the full range of farming and seasonal field techniques inherited their work. Katz notes that these young women's knowledge of their local environment and its resources, put to one side at puberty, is extensive and that this knowledge was drawn upon to build new productive and reproductive relations in much the same way that women have taken control in other global south contexts (see also Warren 1987, 1990). Katz (1998, 136–7) argues that knowledge such as this is not made obsolete by new global economic structures of production and reproduction. Rather, the acquisition and deployment of knowledge underwrites resistance as well as reproduction. The question remains, however, as to the effectiveness of these youths' resistance to globalization with its flexible resources that can easily be moved elsewhere. As suggested by Willis (1981 [1977]), resistance may itself be an element of the reproduction of class relations and work ethics. Indeed, efforts at resistance are perhaps not far removed from a politics of survival.

Institutionalized education and social capital in Baltimore

Patricia Fernández Kelly (1994) argues that there are areas where institutionalized reproduction breaks down to the extent that survival and resistance are sometimes difficult to pull apart. She articulates the story of Towanda who, at 12 years old, is described as "a gorgeous woman-child" living in a poor African-American neighborhood in Baltimore. Just after her fourteenth birthday, Towanda became pregnant for the first time and by 17 she was expecting her second child. Almost completely illiterate, Towanda abandoned school so that she could care for her children.

Fernández Kelly uses Towanda's case to focus on the interplay of cultural and social capital in particular locales (Bourdieu 1984). Cultural and social capital are forms of knowledge that derive from local contexts and are not necessarily tied to formal education but none the less directly impact economic practices. Cultural capital comprises a symbolic repertoire, the meaning of which is learned and used by members of particular social networks. Social capital depends on recipro-cal relations between individuals in the group. Cultural capital represents "ways of talking, acting, modes of style, moving, socializing, forms of knowledge, lan-guage practices and values" (McLaren 1989, 190). To many adults, traits exhib-ited by young people (naiveté, tardiness, politeness, disrespect, honesty, belligerence) are tied up with their body language and sense of dress to suggest a developmental stage or natural qualities when such traits are often culturally inscribed and linked to the social class standing of the individual. Young people's cultural capital is often exhibited through expressive behavior. It differs from their social capital, which depends upon relations of trust, cooperation and reciprocity amongst individuals and groups that may or may not include adults. What is assessed as an exhibition or performance of cultural capital has little to do with stages or essences. An adult focus on rites of passage and natural developmental categories hides the ways that subjectivities are produced as outcomes of class, gender, race and ethnicity and it hides why these outcomes are important for larger scale processes of capital expansion. The question of how young people are placed in society is elaborated through social and cultural capital. Fernández Kelly argues that the form and effects of cultural and social capital are defined by local geographies such as the characteristics of urban space, and other material conditions of lived experience that relate to class, race and gender.

> Because people derive their knowledge from the locations where they live, they also expect that which is probable in their nearby environ-ments, and they recognize as reality that which is defined as such by members of their social network occupying proximate spheres of intimacy.
>
> (Fernández Kelly 1994, 89)

These local geographies are related to Massey's (1993) *power geometries* because they refract and reflect global economic processes, but the process is fragmented along class, race and gender lines to the extent that the meanings and explanations of certain actions and inactions are not immediately evident to the observer. Massey's (1993, 68) metaphor is useful because it implicitly assumes that "the specificity of place also derives from the fact that each place is the focus of a dis-tinct *mixture* of wider and more local social relations and . . . that the juxtaposition of these relations may produce effects that would not have happened otherwise."

Fernández Kelly wonders if Towanda's actions, and those of other girls in impoverished neighborhoods, are meant to deride the mythic notion of American family values. She questions whether Towanda is a passive victim of a situation over which she has no control. In what sense is Towanda resisting racist oedipal family constructions that highlight patriarchal authority? When questioned about her children and an earlier avowal not to get pregnant before finishing school, the young woman replies "some things are just meant to be" (Fernández Kelly 1994, 88). In what sense does this fatalistic attitude reside in a context of impoverished female-headed households receiving government assistance? From this institutional perspective, the burden of poverty diminishes single African-American mothers' ability to supervise their children who mature into a cycle of abandoning school, unemployment, babies and destitution. Economic dependence on public assistance fosters a denigration of paid employment and a continued reliance on government hand-outs. Right-wing conservative policy-makers argue from this perspective that Aid for Families with Dependent Children (AFDC) be cut. Indeed, in some states women who become pregnant while receiving assistance lose their entitlement to additional stipends. Many conservatives believe that welfare programs promote school drop-outs, idle youths, drug addiction, deviance and teenage pregnancies. Fernández Kelly (1994, 95) argues, alternatively, that diverse behavioral patterns are geographic, and that statistical normative patterns of rising teenage pregnancies amongst African-American adolescents do not necessarily imply a deviation from social and cultural norms that are prescribed at the local level. She is particularly concerned that explanations based on culture are really about economics and conceal a neo-classical outlook that suggests young women choose motherhood in order to maximize the benefits that accrue from public assistance. Neither a focus on culture nor economics clarifies the geographic role of social and cultural capital in the transmission of poverty or in the interplay between local economic decisions and global economic processes.

James Coleman (1990) argues that social capital does not arise from individuals but from the relations between individuals. As a consequence of these relations, Robert Putnam (1993) notes that social capital is a moral resource because it involves reciprocity, and the more reciprocity the more useful it is as a resource. If social capital is not used it may diminish as a resource. Using the notion of social capital in a similar way, Sarah Holloway (1998a, 1998b, 2001) is one of the first feminist geographers to question what counts as a child-care resource and who, specifically, needs it. By so doing, she begins the difficult task of problematizing the relations between parents and child-care institutions and engages critically with how child-rearing is constituted as a spatial practice. Holloway's work dovetails with that of another geographer, Isabel Dyck, because it focuses on child-care cultures. Dyck (1996, 126) argues that how women define and negotiate a space for child-care is not merely "local" but revolves around the complex changing daily lives of women (and men). Holloway (1998a, 31) takes this further by recognizing that a "moral geography of mothering" is constituted in a "localized discourse concerned with what is considered right and wrong in the raising of children.'' This moral geography is part of child-care cultures that accumulate objects and signifiers that seem to be desirable and reflect status. There are no deterministic relationships between social and cultural capital, with the former made up by social relations that comprise a network and the latter comprising the dispositions and habits acquired through socialization. Key to both social and cultural capital is that they represent the sum of actual and potential resources that can be mobilized through membership of groups and they have the potential to be transformed into economic capital.

Perhaps the most important limitations on the formation of social and cultural capital are geographic. They encompass notions of proximity and access, and certain norms of inclusion and exclusion. Holloway (2001) conceives of social capital as constituting very real socio-spatial boundaries that affect the lives of ''ordinary'' people everywhere. In particular, she argues that social and cultural capital are embedded in personal social networks to the extent that they (re)produce different experiences and expectations. Her work with young mothers living in Bristol speaks to the kind of moral geographies of mothering that Fernández Kelly reveals as defining Towanda's actions and those of ''women-children'' like her. Parenting qualifies individuals for membership in the adult community and so teenage motherhood is not a deviation from, but a path to, covert cultural norms. Motherhood is a form of cultural capital, a by-product of social capital, because it constitutes a repertoire of symbols that are placed in circulation among a peer group and tapped by individuals to make sense of their experiences. It is a symbol that creates power within a larger structure of domination and extends itself through cricles of mutual support (Fernández Kelly 1994, 100).

Fernández Kelly questions whether Towanda's 10-year-old cousin is choosing public assistance when she articulates a longing to have a child so she can move in with Towanda and they can have their own home. What do children dream about, she asks, when longing for babies of their own? Most of the young women she interviewed had no reason to see education as a path to success; it was, rather, a social arena where they could struggle for self-affirmation. The African-American community and familial network often establishes distrust in educational institutions because it provides children with examples of kith and kin who have experienced maltreatment and discrimination. If children see individuals in their family and community who have been unable to attain an adequate education, who have been discriminated against when seeking a job, or who have otherwise been victimized by institutional racism, their motivation to achieve academically is diminished (Ogbu 1987). Substantial research has shown that ethno-racial awareness develops in children at an early age. Although this awareness may begin with a rudimentary understanding of ethno-racial group and positive feelings of belonging, many authors suggest that as the child experiences prejudice they soon devalue their African-American background. Gomes and Mabry (1991, 166), who review this work, point out that not only are the labels used by African-American children different but their linguistic models are also different because their environment is culturally and experientially different. The movement in the United States from overt to subtle and unconscious forms of racism, as suggested by Nast (2000) in her exposé of the racist oedipal family, poses difficult challenges for African-American children. When these children enter school they are often expected to communicate, interact and learn within a Eurocentric framework which is quite different from what they are used to at home. The subtle message is that Africentric styles of talking, walking, dressing and thinking are not acceptable (Ogbu 1987).

Education was not, for the girls interviewed by Fernández Kelly, a child-centered pedagogy that helps develop a higher sense of reason. Rather, in impoverished neighborhoods where adults often compete with adolescents for meager resources, the school yard becomes an important location for the creation of identity. When other resources (cars, money, jobs) are in limited supply, social capital traverses other avenues including physical prowess, the defense of turfs, tattooing, hairstyling and other corporeal adornments. Images of power are thus embodied when other external representations are unavailable.

Hyams (2000) notes that most of the young Latina women she talked to from poor Los Angeles neighborhoods were convinced that completing or abandoning high school, and how that action relates to their future success, rests on their femininity, bodily comportment and sexual morality. Importantly, she (2000, 635) points out that for most, studenthood was not a generic stage in the life course, but one that was encompassed by larger societal expectations and embedded in local social capital: "It is both in school and through schooling that these young women negotiate gender and sexual identities in their engagement in day-to-day discursive practices and processes." Hyams articulates a space for the young women's sexuality that is juxtaposed between their agency and control by the boundedness of school and cultural institutions. Through what she calls a "spatiality of protection," institutionalized through school dress codes and the prohibition of affectionate behavior, Hyams (2000, 649) argues that young women are subjectivized as simultaneously desirable (vulnerable) and dangerous (desiring). Herein lies a fulcrum upon which it is possible to hang not only sexist but also racist structures of representation. And, importantly, the maturation process is complexly woven into local geographies and it is from these cultural contexts that behavior (and a grounded sense of morality) arise.

An important point about Fernández Kelly's defense of social and cultural capital as local geographies is that they point to a form of morality that is locally embedded, and physically embodied through local actions and inactions, and also bodily appearances. The notion of childhood is complexly woven with local contexts, the marking of bodies and the articulation of sexuality. To say this is not to suggest moral codes and values that differ markedly from larger societal norms. As Fernández Kelly pointedly notes, most of the Baltimore residents she interviewed embrace mainstream American values and identify with notions of hard work, individual achievement and family loyalty. Larger social values, then, are as much part of the cultural repertoire of the rich as they are of the poor. But the social capital that children receive on the basis of family ties may yield meager benefits when relatives are not part of larger networks controlling desirable resources.

Transforming reproductive activities in Zimbabwe

Elsbeth Robson (1996, 2001; cf. Robson and Ansell 2000) is also concerned about reproduction, but in the form of hidden childwork that presupposes a survival initiative where desirable resources are for the most part unavailable. Specifically, her focus is on young carers of HIV/AIDS sufferers in Zimbabwe. In general, Robson (2001) is concerned about children's reproductive labor within the changing spaces of the global economy. She identifies as young carers, children who perform unpaid caring work for ill or disabled relatives in the family home, but points out that to impose the definition "young carer" perhaps pathologizes their activities. Some of the children she works with do not see caring as part of their everyday lives and others would not fit into the definitions of caregiver as prescribed by the global north (Robson and Ansell 2000, 191). None the less, with the AIDS/HIV virus decimating many African populations, there is an increased burden on young people as caregivers however the term is defined. This burden is complicated by larger global economic changes and, specifically, the loan stipulations of the 1980s onwards of the International Monetary Fund and the World Bank that required adjustment to free market principles in over fifty countries. Robson claims that in many African countries, and specifically in Zimbabwe, these structural adjustment policies have hurt children by increasing child labor, mortality, morbidity and illiteracy. In the early 1990s, Zimbabwe's economic structural adjustment program entailed cuts to social services including health and education. It also included the removal of food subsidies and the introduction of health user fees. Trade de-regulation and the privatization of governmental bureaucracy continues in Zimbabwe today at the behest of the IMF and the World Bank. The brief post-independence economic boom of the early 1980s was quickly shadowed by high rates of debt and slow economic growth through the 1990s. Although considered by some to be an adjustment success, the designed macro-economic changes wrought significant constraints on poorer households' access to health, education, income and food. As these burdens of social reproduction forcefully befall women, young people are expected to shoulder some of the burdens of care. This burden is further weighted by the pandemic rise of HIV/AIDS in Africa, which may infect up to 25 percent of the adults in Zimbabwe. Robson's work is some of the first to recognize the labor of young people as care-givers to people with HIV/AIDS.

Through in-depth interviews with young people and with professional NGO social workers and UNICEF workers, Robson elaborates on several stories of carers who "give up their childhood" to help an ailing relative. For the most part, adults made the decision that young people should become carers, often at the expense of the children's education. It is not unusual for Zimbabwean youngsters to expect to be subordinated to their parents' decision-making until they are well into their twenties. Robson notes that age, gender and class are significant factors in determining who becomes a caregiver with poor working- and middle-class girls of around 15 years being the most common workers in this arena. Although one of her informants is a boy she notes that gender is irrelevant in his case because in the poverty of their circumstances there was no one else left to care for his ailing mother. Girls' early socialization in reproductive tasks makes them more likely to become carers than boys. Robson reveals how young people who care for relatives generally lose touch with friends of their own age and find that their experiences as caregivers result in their having little in common with their peers. Other costs include the tiring demands and difficulties of looking after a sick person and the trauma of facing illness and death.

Several of her respondents noted important benefits such as growing maturity, becoming strong and taking on responsibilities that empowered them as household decision-makers. There is also pride in caring and the benefits of a close loving relationship between the caregiver and the care recipient. After caring for a mother with terminal cancer for eleven months, for example, one 17-year-old expressed the ambition to nurse in a cancer ward. Robson points out that the articulation of benefits may be part of strategies for sheer survival, and that social isolation and shouldering responsibilities of these kinds can take huge tolls. That these tolls, and the very existence of full-time child caregivers, are not recognized by many Zimbabwean officials suggests to Robson that this is indeed a hidden form of child labor and that these children will continue their toil unsupported by governmental agencies. She found that members of NGOs and other non-state officials were very critical of the inadequate governmental support for young caregivers.

The larger point that Robson's work highlights revolves around the implications of current global economic restructuring for children's lives as household workers. She illustrates one way that children are involved in and affected by macro-economic processes. Moreover, and importantly, children are not passive victims or recipients of the excesses of global economic restructuring. Robson's stories suggest that although most children do not choose to become caregivers they do not focus on caring as negative and a loss of childhood. Rather, they see it as a learning process and as a conduit to gain maturity and confidence. For many children, caregiving fosters self-esteem and the positive experience of feeling valuable, appreciated and special. From a policy perspective, however, it is clear that most of these young caregivers need financial or material help to alleviate poverty that comes from loss of income (their own and that of the ailing relative) and the cost of looking after a terminally ill patient.

At a global level, industrial restructuring in the global north in conjunction with an aging population diminishes national tax bases that were until recently able to provide health and welfare services. Similarly, countries in the global south are unable to meet the costs of social reproduction as expenditures increase for defense and national debt. As the global south is caught up in the web of flexible accumulation, the resulting widespread economic uncertainty, unemployment, and decreasing public services drastically affect the lives of children. Katz, Fernández Kelly and Robson argue that the price paid for these global changes is the same everywhere because a withdrawal of public welfare funding places the burden of responsibility on private resources and those who can least afford that suffer the most: women, children, minorities, the poor and the working classes. The materially privileged are affected also, as the conditions for economic well-being appear more tenuous and the future of their children less certain. Educational systems respond to prepare children for this uncertainty with increasingly shorter cycles of radical reform in an attempt to engender a sufficiently flexible body of human capital.

Unhinging the local and the indeterminacy of the global child

Ariès' work inspired the idea of childhood as a socially constructed domain, but it also suggested that the creation of modern childhood was historically and geographically specific. Through the early industrial era, the domestic sphere was the site of reproduction, nurturing and disciplining a new labor force that was kept healthy and appropriately educated by local institutions. Contemporary globalization changes all that: now capital has fewer commitments to reproducing any particular labor force or conditions of production. In recent work, Katz (2001) argues that because global capitalism has an aura of placelessness it is able not only to unhinge social reproduction from particular forms of production, but local practices from global participation. The relocation and disinvestment of capital in certain places results in a particular and local unhinging of production and reproduction. As suggested in the above examples, local unhinging causes great stress for communities that are forced to pay more privately for education and health as public coffers are used increasingly to maintain current and attract new investment. Globalization changes the nature of childhood rather than making it disappear. That said, the work of Postman (1982) and others on vanishing children challenges us to explore the nature of current assaults on modern notions of childhood and to highlight the complexly mediated relations between processes of globalization and the pretensions we weave around children's bodies and minds. Capitalism is not committed now to tying down the nature of childhood and so the constitution of a globalized child is unsettled.

The examples raised by Katz, Fernández Kelly and Robson suggest transformation in local networks of capital but they are also stories that reflect the effects of global economic restructuring on the urban poor. In the global north, urban disinvestment, the movement of capital investment to peripheral sites, social disenfranchisement and the deskilling of the urban poor contribute to continued marginalization of neighborhoods like Towanda's. One of Fernández Kelly's (1994, 93) informants sums up these changes by pointing out that their neighborhood

> was always poor. The difference is that in those days we was working poor; now many people are just poor. I worked in the mill and others were working people like myself. When those jobs were lost they had nothing; they had to go on public assistance. . . . And then, when the businesses closed, drugs took over.

Towanda's abandonment of formal education, and her living with pregnancy and poverty in an affirmative and empowering way is woven through with this larger story of global economic restructuring. Katz and Robson also articulate the ways that global processes shape young people's home lives and structure their wider life experiences. Their work in Zimbabwe, Sudan and New York reflect the global diversity of young people's experiences of childhood, home, family and work. Suggested from these studies is a need to be cautious about how childhood is constructed and what meanings are inferred by productive and reproductive labor. If the nature of childhood is indeterminate and circuits of capital are now unhinged from the local, then we need to take seriously the possibility that we are currently witnessing a profound restructuring of the child to fit the needs of global capital.

The central argument here is that children living in the de-industrialized north and the structurally adjusted south find themselves in remarkably similar situations in terms of geographies of reproduction. The important caveat is that ''the translation of these values into action is shaped by the tangible milieu that encircles'' (Fernández Kelly 1994, 89). Put simply and geographically, knowledge, maturation and morality originate and develop in circumscribed spaces.

The separation of children from adults, the taking on of responsibilities and the enactment of rites of passage – the seeming nature of childhood – are all complexly tied with local contexts and larger global transformations. The message from the examples in this chapter is that local geographies structure the meaning of behavior as much as disembodied notions of morality. Indeed, for a message of morality to diffuse throughout a larger society, it must be personalized "through iterative transactions in proximate spheres of intimacy" (Fernández Kelly 1994, 99). It may be argued, then, that a focus on abstract notions of morality exacerbates, to a large extent, the moral assault on young people. The shifting boundaries of this assault need to be understood if we are to better apprehend the resistances that enable young people to jump scale from the local to the global. I argue in the next chapter that the indeterminacy of childhood as a global construct and a moral assault is highlighted further when contemporary behaviors of children are unchildlike.

Notes

1 In his intensive study of the early years of the Plymouth colony John Demos (1970) notes that fathers tutored all their children in moral values from about the age of 3. Sometimes the control went further than mere tutoring as the child was saved by not sparing the rod.

2 Of course, women have not always been at home with children. The factory and labor laws of this time removed women from workplaces or reduced the number of hours they could work to the extent that it may be argued that they were infantilized. Women were not only pushed into the private sphere; at some levels they are considered the same as children.

3 It is, of course, arguable whether a space of family and community harmony ever existed, as Stephanie Coontz points out in *The Way We Never Were* (1992).

6

DESTINED TO SUFFER THE MOST

The last century's focus on children as an exclusive category of existence resulted in the creation of spaces that are designed to regulate behavior and offer the interpretations, prohibitions and examples of adults. In the previous chapter, I suggested that it is important to understand how these spaces of reproduction are changing with disinvestment and restructuring, and what it is that remains for young people. Put simply, childhood is the complex outcome of structural social and economic shifts, and other material transformations. But it is important to understand also that theorizations of children's bodies, subjectivities and maturations are suspect if they are not embedded in local contexts. On the one hand, in a world of shifting values and challenged boundaries there is a concomitant retrenchment and boundary fixing around bodies, sexualities, families, nationalities and moralities. To make sense of the processes of a once localized and now globally extended capitalist order, we need to explore how certain objects come to be invested with seeming natural borders and objective solidarity. On the other hand, there is a need to take heed of the fluidity and constructedness of local geographies in and through children, and the hybridity of their embodied and embedded selves. At some point, the naturalized frames of social life break open as deviant and frequently contradictory processes emerge. At one level, there is nothing new in what I am saying. Discursive assertions of deviance, the elasticity of identity and its local embeddedness are, I think, quite familiar. The point I want to make with this chapter is that these assertions of social constructedness do not, on their own, assuage the current moral assault on young people, their actions and their spaces. Here, I outline aspects of this assault and call not only for acceptance of children's identities, minds and bodies as fragmented and fluid, but also for notions of justice and rights that are not theorized out of a mechanistic notion of discrete individuals who possess essential characteristics and forms. Children's notions of identity, like those of adults, are fragmented and artificial, belying any notion of organic unity or mechanistic wholeness. If identities are not constructed through hegemonic norms, then there is always the potential for the transformatory, the radical, the subversive. Similarly, if justice is not constructed under purest regimes, then there is always the potential for tolerance and liberation.

The crucial task of researchers now is to develop more powerful interpretations of the role of children in the structures of a globalized modernity, and the processes through which once localized Western constructions are exported around the world. In this chapter, I focus on the global political, economic and cultural transformations that currently render children as dangerous and con- tested. To understand the moral assault on young people more fully, I argue, there is a need to understand the transformation of childhood from the vantage of the disillusion of public and private spaces that, among other things discussed in previous chapters, highlights unchildlike behavior. The outcry against child labor in the global south and the reaction to child/youth violence in public places, for example, raises important questions about what constitutes the unchildlike. These questions reside in a moral geography and, as such, establish a conduit for my comments in the final chapter on children's rights and the possibility of new forms of justice.

The unchildlike child

On April 20, 1997, Eric Harris, 18, and Dylan Klebold, 17, strolled into their high school in suburban Denver and began a four-hour killing spree that left fourteen students and a teacher dead. Then, they turned their guns on themselves. Much ranting and spouting off by religious leaders, journalists, politicians and policy-makers inevitably followed as adults tried to make sense of something from which little sense could be derived. Blame was apportioned to ineffective school security systems, the National Rifle Association (NRA), the existential angst of loners, the increasing use of gratuitous violence in movies, television and video games, and the penchant for spectacle in local and global media. It made some people feel better to identify a target upon which to unleash out- rage, frustration and anger. Some condemned changing sets of societal issues including failing community enrichment, a seeming decline in family values, and the rise of racism and neo-Nazi movements.[1] Some consoled themselves with the knowledge that Harris and Klebold were aberrations, one-of-a-kind monsters, while others felt uneasy that there was something larger behind it all. They sensed a systemic collapse into a crazed social imaginary where the worst of our being is what is real. Some feared that attempts to alter that reality by going after the NRA and Hollywood were at best Band-Aids.

Four years earlier the death of toddler Jamie Bulger at the hands of two 10-year-olds, Jon Venables and Robert Thompson, stimulated a similar moral panic and media frenzy in the United Kingdom. Bulger was abducted from a public shopping mall by Venables and Thompson (a closed-circuit television captured them leading the toddler away) and taken to a vacant lot nearly two and a half miles away where he was tortured sexually and beaten to death. Through the spectacle of media filters, there followed an unleashing of public fears that British society was changing for the worst and a whole generation of children were growing up warped and dysfunctional. I do not believe that this is the case at all, but it is not my intention here to offer any kind of salve for those kinds of fears. Rather, I want to try to contextualize unchildlike behaviors in terms of the material transformations I outlined in the last chapter. I also want to speak to what might constitute current social constructions of kids and teens. Gill Valentine (1996) notes that the media coverage after the death of Jamie Bulger and leading up to the court case headlined "the end of childhood" and the "death of innocence." One interesting aspect of the trial case is that the judge overruled prior English law that established children under the age of 14 as not responsible for their own actions, further fueling the claims of journalists about the demise of the innocence of childhood.

If the construction of innocence no longer holds, perhaps the actions of Venables, Thompson, Klebold and Harris are appropriately labeled wild? If the view of children through much of the nineteenth and twentieth centuries was articulated through notions of their innocence, aberrant cases such as these are definitely part of the demonization today of "other people's kids" if not a whole generation. The important connection that Valentine (1996, 596) makes is that this construction of young people is part of a moral panic that reflects a larger set of social transformations. These include changes in families, the unhinging of production from reproduction, the disillusion of the public/private nexus, increasing corporate surveillance of individuals and the commodification of large portions of the private sphere. In this world of shifting values and challenged borders and a concomitant guarding of boundaries around the body, sex roles, the family, ethnic purity and so forth, there is increased anger at young people who cannot or will not fulfill their expected roles. It seems clear to me that the construction of childhood is complexly mixed up in these larger shifts and retrenchments to the extent that there is a moral assault on young people.

Violent events like the Columbine school shootings and the Jamie Bulger case (and there are others) seem to ascribe to young people an independence, autonomy and self-interest that is irreconcilable with the nature of childhood as prescribed through most of the nineteenth and twentieth centuries. Alison Diduck (1999, 127) suggests the contemporary disruption in boundaries of meaning around what constitutes childhood is engendered by increasing numbers of children acting in unchildlike ways, of which violence is just one. What constitutes unchildlike behavior is more often than not behavior beyond the comprehension and control of adults. The ensuing moral crisis does not necessarily herald the death of childhood but rather, according to Diduck, its (and society's) transformation. The sense of a so-called disappearance of childhood is, in actuality, about the loss of a stable, natural foundation for social life that is clearly linked not only to laments over the lost innocence of childhood, but also to a growing anger at, and fear of, children. Diduck outlines several examples of the breach of boundaries between childhood, youth and adulthood that I would like to extend and elaborate upon from a geographic perspective that embraces my previous discussion of the globalization of childhood.

Children as economic consumers and commodified packages

The 2001 *Guinness Book of World Records* heralds Gameboy's *Pokémon: The Yellow Version* as the most anticipated and fastest selling software program of all time, selling eight million copies worldwide within ten days of its release in September 1998, outstripping its rival Microsoft's *Windows 95* by several million. It seems that one possible breach in the global moral fabric of adult awareness is children's increasing activity and power in the market. This is evident in the global north

Figure 6.1 The rapid rise of *Pokémon* as a commodified kid cult and global marketing phenomenon highlights children as increasingly the most important economic consumers. These kids are swapping cards at a *Pokémon* convention

where children consume directly and are particularly susceptible to television and video advertising. In addition, they are able to influence household consumption patterns in a myriad ways to the extent that it is not difficult to reconcile the identity "child" with that of "economic consumer" (Diduck 1999, 128). Not only do they lack childhood as it is constituted and disrupt ideas of children as innocents, but their economic savvy puts them in competition with adults in an arena where resources are sometimes scarce. If their actions are tolerated or even lauded by adults, it is usually through condescension that anticipates an imminent fall.

A further issue here relates to the content of young people's consumption, particularly when it is removed or distanced in some way from adult control. The articulation of young people's television viewing, internet communications, video-game playing and music listening as a moral panic is a regular occurrence in the press and on radio and television. In an intensive survey of internet use by children, Gill Valentine and her colleagues (2000, 159) sum up these moral panics in terms of adult concerns about the relations between home-based private moralities and potentially corrupting and invasive public cyberspaces:

> The off-line world – particularly the home – is imagined as a space of childhood innocence where children are assumed to have no access to pornography or unsuitable materials. Instead these worlds are imagined to be contained within an "on-line" world which contaminates the so-called "real" world by breaching the sanctuary of the home, invading and polluting it with sexually explicit images and "dangerous information."

The invasion of public commercial interests into the home is sometimes seen as an invasion of privacy and a potentially corrupting influence on so-called family values. In most cases, argues David Buckingham (1994), media and capital coercion is conceived as corrupting and all powerful whereas young people are constructed as innocents, dupes and in need of protection. This assertion is, of course, complicated by the notion of young people as active agents. Valentine and her colleagues (2000, 170; see also Valentine and Holloway 2001) note that these moral panics are adultist in that they make assumptions about children's practices: "Children are supposed to be less able than adults to distinguish between suitable and unsuitable sites; less able than adults to handle on-line dangers; and less capable than adults to use technology sensibly without becoming techno-addicts and social loners."

One of the consequences of portable music players and downloadable music files for young people is that much of what they listen to can be hidden from adult ears. Focusing on the consumption of kd lang's music, Valentine (1995) demonstrates its power to articulate sexual identities, subversive communities, and the production of queer space. "The sense of community," writes Valentine (1995, 479), is built upon "a fleeting symbolic fiction of contrived intimacy and unity . . . the power of the fluid dynamism of music in the street is its ability aurally to conjure up or suggest something or *somewhere* that goes beyond the here and now." Music and video games constitute a large part of young people's cultural capital and, for the most part, it is a form of capital that is less accessible to, and thus less controllable by, adults.

A related aspect of children and the market is the way they are commodified packages for market exploitation. This kind of designation applies to all kinds of niche marketing but the way children's care and recreation are increasingly contracted out of the private sphere of the family is of some concern. John McKendrick and his colleagues (2000), for example, note that while children's birthdays in the United Kingdom were traditionally celebrated in the home, parties are increasingly being contracted out to commercial institutions. Herbert Hengst (1987) argues forcefully that the current "liquidation of childhood" is primarily because the bulk of care for children in the global north is now part of the public sphere. The similarity of employment conditions for child-care professionals and all other productive activities is, for him, a prime example of a homogenization of market principles concerned with realigning for greater profit the relations between productive and reproductive activities. Noting this, David Oldman (1994) cautions that because capital still controls productive activities, children will always exist as a hidden "class" to be exploited. The proliferation of commercial child-care in the global north brings fewer unstructured contexts for children and more segregated urban spaces. The disassociation of parents from increasingly younger children and the creation of commodified childhood experiences also raises questions of the changing interdependencies between children and their parents whereby parents have less control of their children's day-to-day experiences. McKendrick and his colleagues (2000, 113) argue, alternatively, that the shift of children's activities (such as birthday parties) away from the home sphere to commercialized spaces represents an extension of children's environments: "Commercial play spaces ascribe the right to play in domains and locales which were hitherto the preserve of adults." But at what price? Constraints on access to commercial facilities or child-care may result in families struggling privately and imploding in upon themselves and, for those who can afford it, segmented day-care landscapes may erode relations between adults and children because they precipitate an understanding of individual children and adults and their relations as segmented, essentialist, exclusive and controlled. Day-care and commercial play landscapes segmented by class and race erode well-being as surely as walled and gated communities because they negate the possibility of experiencing difference and diversity (cf. Aitken 2000b).

Media moralities and protected consumer spaces

The headlines are as unsettling as they are unrelenting: Two Children Die in Parade; Teenager Charged in Shooting; Children Need Advice From Peers Against Drugs; Want to Retire Early? Then Don't Have Kids; Child Molester Arrested on School Bus; Kids Who Kill.

(Shaw 2000)

At the same time that children become an integral part of the commodified package of capitalism, specialized spaces are constructed and their contents consumed. Today notions of "stranger danger" and the "corrupting public" suggest that supposed safe havens – home, schools and some commercially secure environments – are the only seemingly proper places for children. Young people are increasingly confined to acceptable "islands" by adults and are thus spatially outlawed from society (Qvortrup *et al.* 1994, 63). Terror tales of child abuse and molestation, incest and murder inundate media images. There is drama and voyeurism to these spectacles, but there is also an appeal that helps substantiate the moral certainty of adult viewers – they are not the abusers in question. The kind of "non-critical-just-the-facts" reporting, sensationalism and camcorder realism that suffuses images of stricken parents and assaulted children provides a reality voyeurism through which some public streets, and the people to whom awful kinds of things happen, are viewed. The object is to plunge the viewer into direct and immediate emotional involvement without any form of analysis, imaginative detachment, contemplation or reflection. To participate in the experience of spectacle is to embrace these images in a visceral, sense-oriented and unproblematic way:

> These images no longer mediate effectively between our private worlds and the public worlds of the city. There is no longer a space of representation, no longer a transitional space, an analytic space, a dream space. There is consequently no space for exploring and transforming the relationship between the inner self and the object world. The dream space of cinema has given way to what we may see as a worry space.
>
> (Robins 1996, 143)

Media representations of this kind are clearly symptomatic of the disillusion of the public and the private that impacts so heavily on children's day-to-day lives. Currently, society worries about an immediate crisis around the treatment of children, despite a recognition that any visible increases in child abuse reside in the reporting rather than the occurrence itself. The message is that abuse and neglect rob children of their innocence, the very stuff of their childhood.

Of course, damaged children should provoke a sense of outrage but I am arguing that this sense is contrived through spectacles and images that are morally and geographically charged.

A direct result of these spectacles of fear is the rise in North America of commercial chains like *The Discovery Zone* (where "a child can be a child"), *Chuck E. Cheese's* and *Kindercare* that offer secure environments. In the United Kingdom, commercial playgrounds are associated with already established family restaurants and pubs, where play equipment is packaged and themed under names such as *Alphabet Zoo, Jungle Bungle* and *Wacky Warehouse* (McKendrick *et al.* 2000, 101). Youngsters eat, learn, play games and tumble around climbing structures and slides while their caregivers relax in the knowledge that their children are safe from the perils of the street. Susan Davis (1997) argues that these enterprises are a shameless commodification of children's lives, designed and promoted on the basis of fear:

> [These] sites are offered as ways of getting customers out of the house, ways to revive downtowns, ways to create a "street" of activity in hypercommercial space. That all this conviviality is based on a carefully cultivated market is more than ironic at a time when, in less

Figure 6.2 Kid corrals. Commercial play spaces provide safe "cages" where children are sequestered away from dangerous public streets

touristic neighborhoods, minority youth are actively prevented from congregating.

(Davis 1997)

In *The Discovery Zone* and *Chuck E. Cheese's* entrepreneurs create "kid corrals" for safe play where gambling is the most prominent value imparted (if children spend more money playing the games, then there is a chance that they will earn more tokens to redeem for toys). Play areas are often designated into age categories so that rowdy kids don't bowl over toddlers. In *Kindercare* and other child-care centers, caregivers go a little further, accepting science-endorsed theories from child-centered pedagogy that focuses on age-specific developmental outcomes. "Poised for rapid expansion in the next decade in a neighborhood near you," at the time of writing *Kindercare* centers employed 23,000 staff in 1,149 facilities in the US and UK, caring for 126,000 children.[2] The children – "learning every day and loving every minute" – are segregated by age into very specific rooms where educational toys and specially trained professionals articulate Piagetian norms.

While middle-class families in the global north implode in on themselves because of media-generated fears of streets, public coherency is wrenched out of children's lives as they are ferried, metaphorically and literally, from one disjointed social world to another. As toddlers, kids and youths grope for meaning from their kid corral experiences, planners zone secure suburban areas and developers build gated residential citadels for those who can afford them. As Bunge (1977, 74) points out: "It is found that rich children are raised in such palatial estates that they are lonely. . . . Scattering children might prove as bad as battering them, and the poor do not have exclusive rights to child abuse." Unfortunately, emotional and physical abuse is the only central coherency of some young private lives. Some cultural critics argue that when combined with various media influences, these abuses exacerbate the moral decline of a generation.

One of these critics, former West Point psychologist, David Grossman (1998), makes the argument that violent video games, movies and music actually cause violence amongst teenagers. Grossman blames violent media for the increase in aggravated assaults by teenagers over the last thirty-five years but fails to note that in the United States these reached a peak in 1992 and then declined sharply. In a recent editorial, Mike Males (2000) points this out and notes that some of the most violent video games appeared in the last decade (*Mortal Kombat* in 1992, *Doom* in 1993, *Quake* in 1996). Any one disturbed teen might be incited to violence through *Mortal Kombat*, but it is an empirical reach to suggest with Grossman that a whole generation is affected. The Bible for that matter might incite any one disturbed teen to violence. In California, as gangsta-rap music, R-rated movies

Figure 6.3 Kindercare – the most rapidly expanding child-care facility in the world –
advertises a safe educational environment for pre-school children

and internet patronage increased in the 1990s, teenage murder rates dropped
60 percent. It seems clear, then, that censorship and other policies to restrict
youths' access to certain kinds of media because they promote violence are
part of an unfounded assault based on problematic evidence.

What is evident is that while adolescents increasingly consume music, video and
the internet, as part of their cultural capital, as a way to separate themselves from
hegemonic norms, their younger siblings are increasingly institutionalized out of
the family, away from potential friends in their communities and into commodified child-care. Lower-class inner-city youths are also a target for competitive
consumption, but rather than being sequestered in kid corrals, they are increasingly
surveyed on the streets. The surveillance and incarceration of minority youth in
European and American cities is paralleled by market forces but it also points
to spatial segmentations and a moral crisis around adolescent culture and public
space. This separation of kids and teens from adult spaces and activities in the

global north promotes the peer cultures that seem to terrify adult society. Christine Griffin (1993, 14) points out that this is primarily a "moral panic over the urban poor, which [in North America since the mid-nineteenth century] focused on young working-class, immigrant and African-Americans, particularly young men in street gangs." Where cities are unable to provide adequate housing, decent schools, safe parks or streets, young people have responded by forming their own communities which offer protection, a "family," and a livelihood. These benefits are, of course, overshadowed by a subversion of other societal values that are destructive (including substance abuse, sexual promiscuity and a general disregard for life). As such, these communities engender fear in the dominant culture, but not because of the symptoms of urban disinvestment and economic crisis that produced them. The "problems" are most often placed on the shoulders of the families and children who have to live them everyday and the causes are often forced onto the marginalized youth themselves and the supposed breakdown of the family. This side-steps the more crucial, and complex, problem of understanding this situation as a form of contestation wherein young people are responding to the oppressive and castigating dominant culture of late capitalism. Sometimes this response is rage and fury over what Katz (1998) calls the "eroded ecology" – derelict buildings, wastelands of unemployment and trashed resources – of public spaces. The rage sometimes precipitates violence.

Critical pedagogist, Peter McLaren (1989, 1995), argues that the eroded ecologies of youth result in young people, as a collective commodity, inhabiting a media generated community that finds form in slogans, signs, sound bites, head-lines and extreme games. Ideals and images are now detached from places, their rootedness in stable and agreed-upon meanings and associations. Complicit with this disembeddedness, the media provide hollow signifiers focusing on per-formance without value. Outlooks for teens in particular, McLaren argues, are increasingly without hope. It is possible to be a healthy and creative teenager in the United States and yet have a life expectancy of eighteen years. If you happen to be African-American and live in certain neighborhoods in Los Angeles the odds are worse. The moral assault is perpetrated by the center as a way of deflecting the core of societal decay onto the shoulders of those least able to object or resist. It legitimizes and celebrates certain constructions of knowledge and reality while others are bracketed, broken or cast aside as irrelevant. For example, in many schools in the United States curricula focused on reading, writ-ing, science and mathematics are favored because of a perceived need to bring basic "standards" back to American education. McLaren (1989, 169) explains this by elaborating links between the needs of corporate investors in big business to compete on world markets and the imperatives of the new reform movement to bring "excellence" back to American schools. The problem with morality is that it is often conflated with "standards" and "excellence," but who defines those standards and what is required for excellence? Whose morality are we talking about?

McLaren and Rhonda Hammer (1996, 103) argue that these processes are not just about commodification, or adults and their moral panics, because the violence is real. In and of itself, the violence has become a floating and potent signifier:

> What *is* eventful for many of today's youth is surviving the weekend *without* being assaulted, maimed, or murdered. Violence has become one of the few secure ways of stabilizing identity in a postmodern culture, one of the few refuges left to escape the even greater horror of living new, radically decentered forms of subjectivity brought on by the reorganization of capitalism and its marriage to the media.
>
> (McLaren and Hammer 1996, 103)

McLaren's arguments complexly relate to the way cultural capital for young people in the global north is attained primarily through competitive consumption and increased social segmentation. This increasing segmentation of children's formative years is symptomatic of the general segmentation of post-industrial Western society and it is not difficult to argue that children of different gender, class or racial backgrounds are being socialized in different ways that reflect the needs of advanced capitalism. But the responses of children to these segmented spaces and norms are not necessarily ones of acquiescence.

Peer cultures and public space

> Inasmuch as adolescents are unable to challenge either the dominant system's imperious architecture or its deployment of signs, it is only by way of revolt that they have any prospect of recovering the world of differences.
>
> (Lefebvre 1991, 50)

Another breach in the moral fabric, then, is the unchildlike ways young people are engaging with what might be termed public space. Ariès (1962, 391) saw the street, the pub and the café as part of a public sphere of sociability that did not exist as fashionable meeting places prior to modern times. But by the late twentieth century, the street was transformed by bourgeois notions of consumption from a multipurpose space for all groups and classes into space that required stricter control and regulation (Friedberg 1993; Fyfe 1998). Valentine (1996, 591) suggests that "the moral panic of the 1990s" around the violent actions of children, particularly though not exclusively teenagers, is in actuality about "uncontrollability and . . . [the] threat to adult hegemony in everyday spaces."

Some of Valentine's (1997a, 1997b, 1997c) work considers how children and parents conceive of and negotiate potential risks from public spaces in urban and rural contexts. She is concerned with local parenting cultures of what constitutes "good parenting" and how children resist rules that derive from these cultures. It is clear from this and other work that quite often adult moral panics over the kinds of environments their children are likely to grow up in, colors residential preferences and, increasingly, how public space is constructed and policed (cf. Aitken 1998, 2000b).

In California, as elsewhere in the global north, implementation of youth curfews, driving restrictions and similar crackdowns are usually followed by claims of huge reductions in crime, drugs and traffic deaths (cf. Valentine 1996). In California, recent data suggest that these claims are, for the most part, unfounded. Police in Monrovia credited a 1994 school-day curfew with cutting daytime burglaries and thefts by 50 to 60 percent. The press trumpeted Monrovia as California's "small-town success" and other cities in the state rushed to impose curfews that allowed youths out in public only a few hours of most days. In a lawsuit deposition four years later that received almost no publicity, a Monrovia police statistician admitted that the earlier correlations were spurious. Indeed, crime in Monrovia had dropped, but no more than in neighboring cities without curfews, and the biggest declines occurred at times when the curfew was not in place. A systematic statewide analysis from 1994 to 1997 found no substantial decline in crime for cities with curfews when compared with cities that gave teenagers unrestricted access to public spaces, and San Francisco, which abolished its curfew in 1992, showed California's biggest drops in violent crime, juvenile homicides and teenage deaths from murder, firearms and other violence (Males 1997).

Tim Lucas (1998) argues that juvenile crime, gangs, youth curfews, moral panics and their geographies are closely connected. Events at the local scale of his investigation in Santa Cruz, California – the formation of gangs, tagging, and other youth practices – are conflated with wider social processes that provided disembedded interpretations. In particular, he notes that the presentation of the Beach Flats in Santa Cruz as a site of disorder and deviancy is informed by larger debates about exclusive suburban estates and gated communities as sites of homogeneity and order. In the desiccated moral landscape of Southern California, gated communities enclose and protect their residents from the perceived gang-related problems of urban and suburban space. The result is a finer and finer sorting of the social fabric. Lucas argues that the current concern about gangs across the United States is a racialized panic that wants to contain and/or bound the migration of ethnic gangs from Los Angeles. In the words of Santa Cruz's director of the County Criminal Justice Panel, "current resources and approaches could be overwhelmed if emerging gang problems 'spill over' from the more heavily impacted areas of the state'' (quoted in Lucas 1998, 156). Unrestrained and undeveloped by the ameliorating effects of institutions ranging from ''whole'' families and good schools, the values of gang-bangers are perverted and twisted. Young people roaming uncontrolled in public urban spaces are represented as malicious predators, the embodiment of dangerous (wild) natural forces, unharnessed to social well-being.

Of course, the evidence may be turned to suggest resistance and young people taking back the streets. Often, this resistance is found in lower-class neighborhoods and is heavily monitored by authorities, but some argue that the monitoring of youth activities is near universal in some cities. Ralph Saunders (1999), for example, notes how a neighborhood policing program in Boston relies upon surveillance and monitoring by local residents. In one case, two community service officers investigated complaints by a ''civilian'' crime watch team of gang activities in a relatively quiet park in an upper-income neighborhood. The only descriptions of the activities were that they occurred on weekend nights and were perpetrated by people who were ''young'' and ''male.'' The community officers contacted local beat officers who, two weeks later, apprehended four white teenagers drinking beer and smoking cigarettes. The youths were brought to the local police station where, because none of the youths had outstanding warrants or past offenses, their parents were called to come and pick them up. A walking beat patrol was added to the area, and at the next crime watch meeting the community service officers announced that the situation was under control.

An important argument suggests that despite the commercialization and segmentation of young lives, for a substantial residual of young people, the ''street'' is an important part of their day-to-day activities (Ruddick 1995; Matthews et al. 2000; Skelton 2000). Hugh Matthews and his colleagues (2000), in a study of children's use of public space in impoverished housing estates in the United Kingdom, found that children as young as 5 years old relied on outdoor spaces during their free time. Streets provide ''semi-autonomous space or the 'stage' where young people were able to play out their social life, largely unfettered by adults,'' and they offer different opportunities for boys and girls (Matthews et al. 2000, 76). Griffin (1985, 1993) was one of the first to challenge the notion of the street as a universal male domain and youth gangs as exuding an extreme form of hyper-masculinity. Tracey Skelton (2000, 96) reaches similar conclusions, noting that although girls in her study of a low-income neighborhood in Wales spend a lot of time on the street, they also used the public space of a local youth club, Penycraig, to escape the confines of the home and the threats of street life:

> It might not be all these teenage girls would like it to be, but it is a place
> for them, somewhere they feel central and important. It provides an

alternative to some of the stifling domestic settings and the contested spaces of the streets. Without a place like Penycraig, these girls would spend most of their everynight lives on the street. Being on the street brings them into much more regular contact with activities which the Penycraig space allows them choices to keep away from: negative youth activities such as car theft, joy riding, drug dealing and violence.

The point about Skelton's work is that it demonstrates that some teenage girls actively work at being in public spaces, at "hanging around" and elaborating a vibrant cultural capital based on the streets. Importantly, they maintain a "geographical presence" through the social capital generated by friendships and close networks.

There is a small but growing cadre of research that focuses not only on the generation of street-based cultural and social capital by kids and teens, but also links their actions to larger global political processes. Sue Ruddick's (1995) *Young and Homeless in Hollywood* recounts stories of marginalized groups of homeless adolescents who, at times, develop identity politics around the creation of particular places and notions of community in Hollywood. She re-conceptualizes the geography of urban homelessness by pointing out the importance of apodictic spaces wherein the behaviors of the homeless are naturalized and reconfirmed. In a discussion of the evolution of a self-contained punk street culture from 1970 through 1990, Ruddick reveals a series of interesting strategic territorial battles in which the punks have the upper hand, challenging the pre-given meanings of space. In particular, they squat in derelict mansions and apartments in downtown Los Angeles. Importantly, it is not until the decline of punk culture and the rise of a street focus service industry for homeless teens that institutional and city powers are able once more to exercise their control through spaces in between what Ruddick calls the front region of service provision and the back region of the street.[3] Interestingly, she points out that it was the street orientation of the service providers to the homeless youth that gave the 1990s Hollywood urban revitalization plan its focus. Ruddick (1995, 189) indicts past homeless research and policy for prescribing spaces of power wherein "a monologue of meaning flowed from the new post-industrial spaces, marginalizing the homeless further in the process, with precious little backtalk." Her arguments dovetail with Neil Smith's (1996) focus on gentrification as a process of globalization that conspires against minorities, the working poor and the homeless in what he calls the emerging revanchist city with attendant "policies of revenge."

Ruddick wants to access young people's resistances to adult control and she achieves this goal to the extent that she shows how youths confront (and sometimes successfully contest) marginality through occupying public space. As we see throughout Ruddick's narrative, attempts by the adult establishment to thwart the efforts of the punks to create community inevitably fail prior to a global demise of punk culture. She argues that the punks (at least the pre-1980 generation) had remarkable insight into the nature of globalization and capitalist exploitation but, in learning to resist institutional environments and in not establishing the kinds of attitudes and practices that locked them into their class position, they were doomed to be superseded by a more compliant and less resistant group.

Qualities admired in adults such as independence, savvy and wariness construct kids and teens as unchildlike. These qualities are frowned upon and sometimes dealt with severely when they are learnt by children and young people on the street. In order to be turned into a child again, ''the delinquent or street child . . . has much to unlearn'' (Hendrick 1990, 43). Ruddick's story of homeless teens in Los Angeles is similar to the tales of homeless street children elsewhere. Harriott Beazley (2000, 2001) identifies ''geographies of resistance'' in Yogyakarta, Indonesia, where a response to the larger ''spatial apartheid'' of the State empowers street children. Despite their cultural and State sanctioned subordination, the homeless children's appropriation of public spaces for their own survival contributes to the formation of what Beazley calls a ''cultural space.'' Using their private language, Beazley (2000, 196) characterizes the street kids' subculture as *tikyan,* meaning ''a little but enough.'' The children's language is part of their cultural space because it creates solidarity, excluding outsiders who cannot understand. And importantly, for authorities, the new language represents a dangerous mixing of diverse cultural backgrounds. The children live in spaces people are supposed to pass through – railway stations, public toilets – in their movements between socially sanctioned nodes of urban life (homes, offices).

Beazley argues that repressive surveillance systems help marginalize homeless children in Indonesia. For example, all citizens over age 17 must carry an identity card, but this requires items that homeless children usually do not possess such as a birth certificate, a family registration card and a home address. Homeless children under 17 do not carry identity cards but they are supposed to be under the jurisdiction of their parents and if they are not, they officially do not exist. Beazley points out that this non-identity makes these children very susceptible to regimes intent upon their removal from public space (sometimes permanently). Because they do not have family or kinship ties, most homeless children are viewed as socially inferior by Indonesian cultural norms. Street children represent a challenge to the society's moral boundaries as they are enforced by the State. The media often portray these children as deviant and delinquent criminals who were abandoned by their parents and are now creating disorder with their socially undesirable appearance, behaviors, habits and lifestyles. These growing social imaginaries are increasingly accompanied by a ''militarization of the landscape'' leading to a systematic erosion of public space and street children. For example, between 1983 and 1985, in an effort to eradicate crime throughout Indonesia, the bodies of more than 5,000 suspected criminals (mostly youths dubbed ''Gangs of Wild Children'' by the press) were dumped on the street after they were killed by ''mysterious gunmen.'' Only in Yogyakarta was it admitted that the shootings and the killings were part of an official war against crime. More recent ''cleansing campaigns'' have been less murderous but they none the less attempt to displace homeless children from public spaces such as bus terminals and shopping centers. Beazley describes stories of children arrested during these campaigns who were subjected to emotional abuse, hunger, torture and rape while in custody.

Street children in Yogyakarta have been able to subvert and resist this oppression and the domination of public space by recoding it in their own interests and through their everyday practices. They know the places – traffic lights, city parks, public toilets – where access is not rigidly controlled and it is through these places that the children find solidarity and it is here that their identities are constructed, enacted and articulated. Toilets are very masculine spaces, Beazley notes, and compliance with peer norms of masculinity is essential to *tikyan* because friendships, support in hard times, and personal safety are all dependent upon the social capital of a peer group. A sense of belonging contributes in an important way to the shaping of social space at a toilet, and an age hierarchy exists with the boundaries around the space policed by older boys. Beazley focuses particularly on street girls who are not only harassed by the State but also by street boys who often call them *perempuan nakal* (prostitutes) or, more commonly, *rendan* (vagrants wearing make-up). Street girls resist these images by saying they feel *cuek* (could not care less) but they are none the less constrained by both the State and patriarchal hegemonic norms. Some resist Indonesian forms of femininity by smoking in public (something ''not done'' in Indonesia), wearing jeans, men's shirts and cutting their hair short. They also occupy different street spaces such as the city parks and squares. They seldom visit toilets unless invited. These public sites are multiple, shifting and gendered, but they provide spaces through which the children create identity and self-esteem: many relish their abilities to survive an oppressive regime.

Beazley's work suggests that street children are not wholly outside normative, patriarchal, socializing control but they do resist certain adult strictures. Stephens (1995, 12) argues that from certain perspectives, street children may be viewed as integral parts of an emerging order of global capitalism: ''Frequently detained for the nebulous crimes of loitering and vagrancy, these assemblies of young people are most guilty of not conforming to socialized models according to which children are compliant vehicles for the transmission of stable social worlds.''

Child labor

Child labor clearly also evokes an interesting morality play as a potentially unchildlike behavior. During the Enlightenment, Western society discovered the importance of play for children, which was then designated as the antithesis of work. Clearly, labor is a particularly problematic concept. When it is stated that children no longer work what is meant is that they do not endure mandated hours of labor under the control of an employer for which they may or may not receive wages.[4] Although what is described as child labor is complex, it is probable that until the nineteenth and early twentieth centuries the mass of people in Western society took children's productive activities for granted. It is much harder to assess the geographic distribution of other kinds of work. Advocating the regulation of child labor began in earnest in the United States and the United Kingdom around the mid-eighteenth century. Although children took jobs in many different contexts including industrial weaving and, of course, farming, advocacy began when children were used for chimney-sweeping and mining. These jobs affected young children, certainly, and in the case of the chimney-sweeps, work became a visible part of middle-class life.[5] Few commentators at the time sought the abolition of child labor but rather argued that it be regulated and humanized, presenting children as naturally ill-suited to heavy labor and pointing out the ways that working in a mine or as a chimney-sweep stunted natural growth and development.

Sue Roberts (1998) argues that a child as part of the global market economy is seen today as a child without childhood, a child who is robbed of the proper experiences and categories of children; in short, a subject defined by a "lack." That these children lack economic capital is obvious, but Roberts points out further that they also lack cultural capital because they transgress the boundary between how adults and children are constructed. And there is a crass geography to the representation of these children as human subjects. The problematic demarcation of child workers within the "Third World" and child consumers in the "First World" represents a process of delimiting and bounding global space into zones of difference. What part the labor of young people plays in the productive and reproductive activities of their families is hard to assess because of a global diversity of children's experiences that clearly goes beyond a simple north–south divide.

Organizations such as the International Labor Organization (ILO) and the International Program on the Elimination of Child Labor (IPEC), that are concerned to protect children from abuse and exploitation, articulate a distinction between child work and child labor in order to classify what is harmful. The rhetoric is of some interest. Fyfe (1989, 4, quoted in Robson 2001), for example, argues that child labor should be defined as work which impairs the health and development of children whereas child work constitutes all work which detracts from the "essential activities of children, namely leisure, play and education." At the time of this writing, an ILO internet site proclaimed that

> Child labour is a pressing social, economic and human rights issue. As many as 250 million children worldwide are thought to be working, deprived of adequate education, good health and basic freedoms. Individual children pay the highest price, but their countries suffer as well. Sacrificing young people's potential forfeits a nation's capacity to grow and develop.[6]

Children's basic human rights are important but the issues raised by definitions of this kind relate not only to what constitutes adequate education but also how that education relates to a nation's capacity to grow.

Elsbeth Robson (2001) points out that although these discourses are often contradictory and uneven, two main issues ground problems with their application. First, international conventions almost always originate in the global north and are informed by Western norms and, second, they attempt to establish universal standards to highly differentiated global contexts.

Given that the work of children (as caregivers or as part of the informal sector) is often hidden, any serious appraisal of their contribution must impact analyses of economies and societies. Robson (2001) argues that an appraisal of this kind is needed in order to give to children the same kind of political clout that, over the last decade or so, has been gained by women. To some extent, Robson's argument is appealing because she is much more concerned about the current plight of children as laborers than their potential to contribute to national economies. One of the issues that her work raises is that the ILO's current focus on finding employment for parents and guardians does not apply to the contexts of young home-based health-caregivers where the parents are unable to work. She argues that unpaid labor in the household needs to be assessed but it is clear that there needs to be recognition that this *is* work and, irrespective of competing discourses on childhood and work, it significantly contributes to national economies. These young workers need support rather than any complicated ILO agenda to remove them from the workforce.

In some families today, and particularly in the global south, the contribution made by children to household income is undoubtedly important. It is now commonly known that children are essential to productive activities as well as family subsistence in global south contexts (Rodgers and Standing 1981). Children's work includes fuel and food gathering, animal husbandry and herding, and food preparation. Children in these contexts may generate income or they may work with parents and so help to raise household productivity. Alternatively, they can make a significant contribution when they look after younger children to enable parents, especially the mother, to work. The problem with articulating work in this way is that it misses nuances that destabilize the meanings of productive and reproductive labor. Robson (2001) argues that immense problems relate to the consideration of work by children in different social contexts. For example, when a young person leaves home they can affect the family budget quite directly, either positively by no longer requiring sustenance, or negatively by withdrawing their labor. Katz's (1991b) work in Sudan suggests that when young men leave rural areas to find work, young girls out of necessity take up much of the males' local labor, even though it is prohibited by the strictures of *purdah*. Nieuwenhuys (1994) suggests that in contexts such as these, focus on children's work is valuable because through work, children can contest and negotiate childhood.

Importantly, the reasons why children work should be sought from the children themselves. Robson's (2001; Robson and Ansell 2000) investigation of young people in Zimbabwe's home-based health care suggests that there is little choice when adults become dependants because of debilitating diseases. Economic restructuring that removes resources from the public health system precipitates this lack of choice. Robson's respondents articulated a sense of loss when they found themselves removed from school and peer groups but this was partially balanced by a growing sense of self-esteem fostered by their responsibilities:

> The diversity of responses among key informants indicates something of the nature of struggles in Zimbabwe around understandings of childhood and notions of work during childhood. Some construct what children do as not work, i.e. as socialization, part of being Shona/African, something expected in traditional extended families as good and proper behavior in training and preparation for adulthood.
>
> (Robson 2001)

For children in these households, care-giving activities are an everyday feature of reproduction and, as such, their circumstances differ from children elsewhere. This raises important questions about the global distribution of the resources that directly impact the lives of children (Aitken 1994, 12–14).

The issues that distributional justice raise relate to why some children must work to ensure household survival while others can over-consume (Holloway and Valentine 2000b, 10). This question, although important, hides the equally important notion that children's work, wherever it occurs and whatever it is for, is perceived as unchildlike. For example, Morrow (1994) highlights a different form of contesting childhood that may be attributed to children's working to enhance their consumptive power. She argues that some children from more affluent families in the United Kingdom work to enable their participation in new teenage consumer markets. Although some of the children she interviewed saw work as offering them experience and self-confidence, for most it was also an opportunity to purchase items not financed by parents.

Bowlby *et al.* (1998) note that occupational choice in contexts such as these is shaped by individual conceptions of self-identity and preference but these choices are constrained by employers' biases, usually hidden, that relate to gender, race, class and age. Moreover, the transition into paid work takes place in specific social and physical ''spaces'' which strongly influence the experience and its outcome. Bowlby and her colleagues use an extensive study of Pakistani youths in Reading, England, to argue for the importance of social and cultural capital, but they argue that for their young women research participants, some distinctive behaviors and relationships (e.g. Islamic dress standards and restrictions on the times they could be away from home) that provided local support against racism also created barriers when interviewing for jobs or ''fitting in'' if they secured a position. Similarly, from her work with young African-American women in Baltimore, Fernández Kelly (1994, 101) notes that although practices of hair braiding and tressing may be described by friends as ''bad'' and ''nappy,'' they may not know that flamboyant coiffures can curtail the probability of success in job interviews.[7] A radical approach to young people's labor would recognize their work, support their right to work, and campaign for that work to be fair and not exploitative. Unfortunately, the rhetoric surrounding children's labor (like the constitution of childhood in general) is often problematically couched in developmental terms. Within the seeming structured coherence of Western rhetoric, the unchildlike behavior of young people in the global south could be interpreted as local instances and particularities of backwardness and underdevelopment, thus justifying greater efforts to export modern childhood around the world.

The media of the global north periodically produce savage accounts of child exploitation in ''Third World'' sweat shops where sportswear and electronics are manufactured with advertising that contrasts the grim factory conditions of child labor in the global south with the glamorous world of star-endorsed consumption. Roberts (1998, 3) argues that campaigns such as these ''press a geographical imaginary. . . . In the contemporary representations of child labor, the time is now and the space is a segregated global one.'' These diatribes parallel media accounts of child abuse in poor Western neighborhoods. By so doing, the global north's industrial past, the present global south and poor neighborhoods everywhere are highlighted as barbaric and exploitative for children. This suggests a problematic continuum of development that is couched in the rhetoric of contemporary modernization and globalization. It is a rhetoric that enables the ''othering'' of adults in past eras for their cruelty and indifference to children so that we can judge favorably how far we have come today. Are we led to believe that contemporary Western society is much more complex in its moral stance regarding children whereas early modern society, and poor neighborhoods and parts of the global south today, have a warped or a non-existent notion of childhood? Such dichotomous thinking hides complex structural processes whereby settings in the global south adjust to changing conditions of capital that widen and deepen global divisions of labor that directly affect children along with women and minorities.

Rethinking last century's childhood

Ariès' ''history'' drew primarily from sources in France and England, but he claims that given the dramatic changes in our understanding of childhood in Western society as a whole, so also has the experience of children changed. He takes for granted that these changes are integrated into larger social and cultural transformations although he does not elaborate on what these might be. But, importantly for the developmental issues I want to discuss now, there is clearly a value judgment throughout his work that the ways of life disclosed in the ''traditional'' places and periods are preferable to those of here and now. He notes for example, that the pedagogists and moralists who propagated large parts of contemporary change through structured educational discipline did so by overshadowing more humanistic concerns (cf. Ariès 1962, 412). Wilson (1980, 151) critiques the methods employed in *Centuries of Childhood* for their tendency to see in medieval society (and, for what I argue here, parts of the global south) only the absences of modern attitudes (and the global north). In short, these absences define our presence. Ironically, in the same way that Ariès highlights the artifacts of French and English sixteenth-century culture to suggest the absence of childhood, so too can contemporary scholarship that embraces the developmental notions of children and nations unequivocally be critiqued for embracing the moral high ground of a ''normal,'' better present.

A large part of what I have written draws from historically and geographically linked precedents but it is not developmental, and so I want to focus here on the currency of what I am calling critical moral geographies of kids and teens. For all its weaknesses, I argue that Ariès' thesis is epistemologically relevant because his is a present-centered approach that views the past exclusively from the point of view of the present (cf. Kincaid 1992, 62). And so, from these beginnings it becomes relatively easy for me to use this chapter to highlight various institutionalized movements that focus on the care and well-being of children (locally and globally) as evidence for what really constitutes collective progress and development. It is easy to pass judgment on past eras, but there is just as much evidence of cruelty and violence regarding children in our own time and, importantly, in our own place.

Reconsidering the moral assault

I pointed out at the chapter's beginning that the violence to, and by, young people is such an emotive topic that we easily reel into confusion and advocate simple moral imperatives to what is an extremely complex problem. What I want to suggest here is that the problem may well reside in how we represent and imagine children and our adult relations to them, and childhood and the mythical developmental framework within which the concept evolves. At one level, then, my concerns in this chapter revolve around the exploited, fragmented and sometimes violent day-to-day lives of children and youths and how those relate to their search for identity and the kinds of social capital they might attain and the cultural capital they might exhibit. To the extent that Klebold and Harris are seen as aberrations they are abstracted from their local contexts in the same way that the Third World child worker is extracted from his or her particular circumstances. I think there is an important link here to other abstractions around monolithic notions of ''the'' child. Thus the figures embodied in the ''trench-coat mafia'' and the ''Third World child worker,'' while perhaps raising emotions that alternate between acrimony and angst, simultaneously disrupt and reinforce the ways in which socio-spatial processes work to bound and mediate identities. At another level, I am concerned about the institutionalization and commodification of reproduction and its unhinging from modes of production. Within this context, children's worlds are still designed to create individuals who operate in the interests of corporate capital, and whose social function is primarily to sustain and legitimate the status quo. And so, to understand more fully what is going on we need to look simultaneously outward and inward, backwards and forwards, locally and globally, north and south. It is at the dislocated margins where we find the reasons for our own, more centralized, developed, adult locations. The state of crisis is not the exception but the rule, and the violence is borne most egregiously by young people because they are, in reality, closer to the margins where sheering and dislocation are most acute.

There are no single, mono-causal explanations for the violence at Columbine or on Jamie Bulger, but there are similarities between the people and places involved and it is these similarities that suggest, at least to me, a need to focus on critical moral geographies of young people. I am not trying to make sense of individual atrocities perpetrated upon young people, but rather I want to use those contexts as a springboard to understanding more fully the place of children and youths in a larger global society. I can do so because problems in society are more than simple isolated events but, rather, they form part of a relational context between people, places and things. In a sense, then, there is a need to defend ''atypical'' behavior on the grounds that it offers some special insight into larger social patterns and processes. Of course, the work of Michel Foucault was a foundation for the belief that the normal and the abnormal are closely linked. The world of children is not all violence and mayhem but there are no simple unidirectional explanations for development and progress no matter how many politicians and cultural commentators bemoan the demise of family values, the need for educational standards and the ostracizing of youth.

One task might be to examine the ways in which young people talk about values and support their talk with action and inaction. It is important to consider the contexts of inactions (as well as actions) that enable young people to unshackle themselves from adult tutelage and global processing. Inaction is different from indifference, which is about not caring and is immoral because it creates despair. It seems to me that the two boys sitting under the !Vivo Sandino¡ sign on the front cover might understand this quite well. A further aspect of not caring relates to attitudes in the United States to the Sandinista revolution and the emotions garnered by my friend's photograph. This is about larger, more complex issues of justice and is related to the topic of my concluding chapter. Insofar as global proclamations about justice and children's rights are acceptable, it might be argued that children also have rights not to be saved, not to be constrained within exclusionary cultural identities and not to have their bodies and minds appropriated for larger political effect.

Notes

1 See Kobayashi and Peake (2000) for a discussion of the violence at Columbine High School as a racially motivated act.

2 This information was gleaned from an internet advertisement at http://www.kindercare.com.

3 Of particular interest is the way that the monolithic social imaginaries of family and community are used as dynamic strategies by service institutions to sustain relationships between the youth (brothers and sisters) and the service (family) structure (Ruddick 1995, 152). It is interesting that the rhetoric of family life is used to ''socialize'' homeless youth when, for many, it is the dysfunction of society and the failure of the family within society that often precipitates their move to the street.

4 Yet children are not free to do as they please and they do not receive payment for attending school. Indeed, McLaren and Hammer (1996, 105) suggest that school is an extended apprenticeship for entry into capitalism: ''education is treated simply as a subsector of the economy to train students for economic wars with Southeast Asia and Europe.'' Whatever way we look at it, school certainly counts as work in some sense.

5 There were also symbolic ramifications of the ragged soot and coal-dust covered children that pressed heavily on the recent abolition of slavery in the United States.

6 http://www.ILO.org/public/english/standards/relm/ilc/ilc87/rep-i.htm#Progressive elimination of child labour.

7 Within a circle of social capital that is impoverished by a low degree of multiplexity, vulnerability is recast into power by bodily adornment. But what is wrought as an empowering local symbol becomes a symbol that bars access to resources in a larger society.

7

A SPACE TO PLAY AND A TIME
FOR JUSTICE

Lefebvre (1991, 416) suggests that for a human group to be recognized by others, as ''subjects'' they must generate or produce space. The idea of value, he argues, acquires or loses its distinctiveness through confrontation with other ideas about value and ''it is in space, on a worldwide scale'' that this process occurs. Global spaces, then, are an integral part of, and not just a setting or stage for, a morality play wherein the qualities and virtues of being human are contested and compiled, and where difference is accepted or repelled. This is the realm within and through which the identities of children, adolescents and adults are contested and continuously constituted. At one level, young people cannot easily be compiled into a monolithic human grouping but, at another level, they are unique in that adults almost always see the importance of creating spaces for young people but are often loath to let them do so themselves. The last chapter underscored this adultist control of space and spatiality by emphasizing the links between child-centered pedagogy, the increased institutionalization of young people in educational establishments, the constitution of child-care, childwork and child-labor industries, and the removal of young people from public spaces on the one hand, and market-niche expansion and globalization on the other hand. The linking of these issues is complex but I tried to raise them in a way that purposively highlighted scale relations between young people's search for political identity and the larger forums within which community may be found, with a particular emphasis on interrelations between global and local performances.

The moral assault on young people may disembody them as wild animals, but it is also embedded in wider cultural practices that suggest childhood is sacred – a time to play and not to have responsibilities, a time of innocence, a time to study, a time for school and preparation for an adult life of paid employment. But cultural capital is dependent upon physical and social location and so lived reality structures the meaning of behavior, for poor and affluent children alike, to a larger extent than disembodied and disembedded notions of morality (Fernández Kelly 1994, 89). These encircling milieux constitute critical moral geographies of young people that enable a broader focus on a space for play, children's rights and new concepts of justice.

Critical moral geographies

Although I believe in the efficacy of local lived realities, I am none the less concerned about an emerging geographic theory of morality that suggests "the good" exists as something tangible and attainable through "places" (Entrikin 1994; Sack 1999). Proponents of this theory argue that the good is an existent reality that may emerge if places enable the world and its parts to be seen more clearly and in its totality, and if they enhance the value of variety and complexity. I argue that this reification of place as central to the establishment of the good is problematic at several levels but most importantly it does little to elaborate the power of scale and the moral assault on young people. A large part of the last two chapters was about geographies of scale, not as existent realities but as a way to bring together local and global concerns that highlight the morally contested spaces of young people. In the last chapter I sketched some points relating to this moral contest and suggest why a focus on the geographies of everyday life is an important way of unpacking and perhaps addressing the crisis. What I want to do here is look more specifically at what a critical moral geography might look like as a precursor to shedding light on children's rights and justice.

Most *moral* theorists begin with the premise that the world is undergirded by a universal search for the good as an existent reality. This does not necessarily imply that the good is achievable or even capable of being represented but that, at the very least, it exists. In short the good is real but ineffable. Most *critical* theorists, on the other hand, begin with the premise that adults and children alike are not necessarily free to seek the good, but inhabit a world structured around contradictions and asymmetries of power and privilege (McLaren 1989, 166). Part of any critical analysis is to tease out relations of accepted meanings and appearances, and highlight contradictions and conformities with everyday practices. Critical geographies focus on spatial relations and power asymmetries (power geometries) as they may manifest themselves in everyday places. Some moral theorists argue that by so doing critical geographers focus solely on distributional justice and what is needed is an understanding that conceptions of the real and good are presupposed by notions of justice. Places are not just about spatial justice, they argue, but about making us less or more aware of reality (Sack 1999, 26–8). In the study of critical geographies, however, places are seen as sites of power rather than as ineffable realities and, as such, both liberation and domination are complexly connected through local places and from other scales of practice. When related to "the good" and defined as conformity to ideals of right human conduct, questions are raised about who has the power to establish conformity, rights, authority and justice. This power is articulated around sexuality, bodily comportment and image, and it presupposes the most appropriate categories of growth and development. Quite recently, this articulation of power has been complicated with discussions of children's rights to culture.

Children's rights to culture

Part of the contemporary crises of childhood, writes Diduck (1999, 128), "is the Western liberal movement which attributes rights to children, and the consequent ability of children to make rights claims." In the nineteenth century, demands were based on the rights of children to play (as opposed to work) and be children. Today, children's rights are encompassed by assumptions of autonomy and self-consciousness to the extent that children can now demand that law respects their human rights and their voices are heard when decisions about them are made. As the expanding local/global nexus reshapes modernity, how are young people affected? How do children experience, understand and resist the complex cultural politics that circumgyrates around and through their lives?

Figure 7.1. Parents from a local school in Kent, Ohio, protest outside a McDonald's restaurant for their children's right to a safe journey home. The entrance to the McDonald's drive-through window intersects with a sidewalk that leads directly from the school

Perhaps the most significant contribution to this recent focus on children's rights was the United Nations' *Convention on the Rights of the Child* (CRC) in 1989. The CRC specified a series of rights for children, separate from and in addition to the rights of adults, phrased largely in terms of general moral entitlements. Children were granted a wide variety of rights including legal, social and civil rights, but the CRC also reflected considerable international ambiguity over the meaning of childhood and the "political" status of children (Kulynych 1999). To date, the CRC has been ratified by more than 190 countries and, as such, it represents the strongest statement in international law that children should have some independent status (Burman 1994; Matthews and Limb 1999).[1] The CRC emphasized the capacity of children to at least partial autonomy without the supervision of adults. It lays down children's rights to freedom of expression and association, for example, that are enabling, as well as to rights that are protective (Cantwell 1989, cited in Stephens 1995, 35).

Roger Hart (1997, 16) argues that the debate over children's rights spawned by the CRC is important because it encompasses a discussion of fundamental changes in culture and how culture reproduces itself. Unfortunately, by establishing children as different from and, to a degree, independent of adults, interpretations of the law may lean toward simplified and polarized arguments. Hart points out that it does not help to think of "equal rights" in the sense of giving children the power to make decisions equal to that of their care-givers. Children depend upon the caring decisions of adults but it is important that there is an openness to listen to and communicate with children. In this context, Hugh Matthews and Melanie Limb (1999, 79) point to Article 12 of the CRC, which asserts "children's rights to be consulted, heard, listened to and taken seriously, in accordance with their age and maturity." The common wisdom here is that with maturity, children's rights increase and they participate more fully in public life.

Public life, as Jurgen Habermas (1984, 1987a, 1987b, 1989) views it, requires that citizens participate in reasoned communicative action that excludes the apparent disorder of children. Or, as John Rawls (1971) argues, democratic theory begins with the assumption that all citizens are fully cooperating members of society. Both Rawls and Habermas focus on moral development as the prerequisite for citizenship. Rawls' notion of justice, rights and democracy emanates from a citizenry that has a conception of "the good." Children eventually reach a stage of reason and attain a sense of the good but their immaturity is not politically relevant because others act on their behalf and do for them what they would do if they were rational (Rawls 1971, 249). The question that the notion of moral maturity raises relates to the particular developmental trajectory embraced by the CRC. Article 29, for example, asserts that the child's rights to education should presuppose "the development of the child's personality, talents and mental and physical abilities to their fullest potential." As a consequence, children who are not fully developed mentally and physically are excluded from full participation in public discourse, the public sphere and public places.

Ideally, argues Hart (1997, 16), children's rights should be directly related to their capacity to participate in the development and management of environments, which, he suggests, is directly linked to their stage of development. Thus, by age 6 they get access to environments that grow into environmental interests and ecological understandings by ages 10 or 12. It is only at age 14 or older that children develop empathy, morals, or any form of political understanding. It is ironic, then, that developmental psychologists have used Piagetian methods to show that children's abilities to make moral judgments are often grossly underestimated by the linearity implied by sequential stages (Ruffy 1981; Ng 1983). Hart is aware of this and notes that an invocation of any kind of static developmental continuum is problematic if it suggests a mechanistic and linear approach to children's rights. Instead, he points to the need for a "more radical social science [in which] children themselves learn to reflect upon their own conditions, so they can gradually begin to take greater responsibility in creating communities different from the ones they inherited" (Hart 1997, 79).

Hart (1997) goes on to suggest a radical perspective that anticipates children's rights in an adult dominated world. In so doing, he jettisons any pretense of a developmental continuum and, rather, offers a typology of engagement from manipulation through tokenism to participation in the form of "child-initiated shared decisions with adults." In this sense, granting of rights to children over shared decisions should not undermine the legitimate responsibilities of parents and caregivers. But the CRC suggests strongly and problematically that biologically based family relations are more natural and fundamental than other kinds of relations between adults and children. Such a view marginalizes other family forms such as single-parent households, mobile and extended families, gay and lesbian families, and non-kinship-based groups. Bracketing and setting aside the assumptions of the CRC for a moment, it is evident that Hart's work relates, at least in part, to Bunge's (1977, 77) vision of turning power relations on their head through a revolution that installs children as the "privileged class" while still placing responsibility with adults. This whole issue, of course, turns on how responsibility relates to authority, and understanding the distinction between "authoritarian" and "authoritative" supervision. In noting this, Hart argues that adults need to recognize children's rights to appeal against decisions and they need to be able to articulate the wisdom of decisions that affect children's lives: it is not enough to say that this is "the law."

But we can only set aside the implicit assumptions of the CRC about families for a moment. In contrast to the nineteenth century's patriarchal authoritarian private family where the parent not only had the "natural right," but acquired the "political right" to command children, in the democratic family of the CRC the right and authority of parents becomes primarily a natural right rather than a political one. The natural and biological parents' "right" has a special temporary nature in that it dissolves once the child becomes "self-governing" (Elshtain 1990, 57). But when applied to children, the "right" often assumed an autonomous, independent, and very unchildlike "being." The problem with focusing upon the process of individualization and self-governance amongst young people lies with being able to establish the limits of "self-governance" in so far as when and where it occurs. A more critical examination of rights to culture and self-government is needed in a world where more and more children are growing up in complex multi-cultural settings, creating identities that bewilder adults. While many argue that international rights discourses further the best interests of young people, in some contexts these discourses are linked to significant risks to the well-being of children. It might be argued further that children also have rights not to be pigeon-holed with exclusive cultural identities or constrained by the tutelage of well-meaning adults, and not to have their minds and bodies appropriated for larger political intents.

Family, civic and corporate authority

Jean Elshtain (1990, 59) avers that, for children, there is little doubt that parents are powerful and authoritative but she maintains that this is not necessarily an important focus. Parental authority may be abused in ways more insidious than any other form of authority, but unless it exists, parenting itself is impossible. Family authority is imperative for Elshtain within a democratic, pluralistic order precisely because it is not necessarily homologous with the (patriarchal) principles of civil society. In that "the law of the father" continues in civil politics it does not necessarily have to in private politics. A direct hierarchical relationship between the private lives of children in families and the public polity weakens democratic principles. Children, Elshtain asserts, need particular intense relations with adults to help them make distinctions and choices as adults. This raises the importance of the ability of child and family contexts to oppose and contest societal norms as a marker of what works in democracy. As I noted earlier, differences in sexual and racial identities are either not considered at all, or thought to be irrelevant or subsidiary to the "normal" process of child development as inscribed by traditional social science theories. Of late, feminists and social theorists argue for a focus on gender and racial formation and the socialization of children within diverse family settings. What Elshtain is advocating takes this a little further: it is that respect of difference (between parents and children within care-giving groups, and between care-giving groups and society) is crucial for the maintenance of democracy:

The social form best suited to provide children with a trusting, deter-
minate sense of place and ultimately a ''self'' is the family. Indeed, it
is only through identification with concrete others that the child can
later identify with nonfamilial human beings and come to see herself
as a member of a wider human community.

(Elshtain 1990, 60)

Authority for Elshtain, then, rejects any ideal of political or family life that absorbs
all social relations under a single authority principle such as patriarchy. She feels
that the replacement of parents as authorities over children would not result in a
consensual world of children with equal rights and the status of adults, but rather
one in which children become clients of institutionally powerful social bureaucrats
through processes not unlike those described in the last chapter. By no means is
this an advocation for a return of children to the private realm.

David Archard (1993, 168–9), in arguing that children's rights should be
extended, agrees that the authority of parents should not be disrupted, nor is it
necessary to cover all children with all rights currently available to adults. His
''modest collectivist proposal'' advocates that children are brought more into
the public domain but not in terms of the capitalist commodification and exploitation that seems to prescribe children's current experiences. Rather, Archard's
concern is to remove the ''private shades of the family'' to make easier the
monitoring of children's physical and psychological progress so that in some situations and for some children there is ''public and palpable acknowledgement of
their status and worth.'' When children become accepted as fully social persons
and not *others*, it follows that the social positions and locations that children take in
contemporary life and the roles that they are allowed or enabled to take in society
will need theoretical rethinking.

Diduck's (1999, 129) concern with Archard's position is that it pays insufficient
attention to the ''worth that comes from mutual need, interdependence and connection.'' Her arguments go one step further, to suggest that the rights accorded
children within our current system of justice require an unchildlike sense of
autonomy and awareness that crosses the classic bounding not only of what it
means to be a child, but also the bounding of private and public spheres. Such
a transformation suggests that the system of justice needs also to change. In practice, then, not only familial but judicial authority today needs to be transigent and
open to the recognition that children frequently have something to teach: transformations of social structure flow up as well as down. The problem with the
CRC is that it disempowers the former transformation. Stephens (1995, 39)
pointedly notes that the CRC establishes rights for the child that are first and foremost couched in a burgeoning international modernist culture, then to identity
(conceived in individual, familial and nationalist terms), and lastly, in special

cases, to minority and indigenous culture. In what follows, I try to move theoriz-ing past the especially tenacious notion of "universal children's rights to culture" by putting the term at risk in a truly playful way.

Play and privacy

There is an insidious irony in how we deal with children's play and their rights to justice. For the most part, young people as key sites of subversion and resistance are foreclosed upon while at the same time international conventions and scholars consider them as key sites in which politics and "the political" are centered. The irony highlights a disturbingly adultist project. What is missing is a consideration of children as playful, active, interpretative subjects, which at the same time understands that they are not autonomous subjects who are always dealt with justly. Keeping in mind that there is no universal form of play any more than there is a single, monolithic children's culture, it is worthwhile considering a space of justice that elaborates play and takes it seriously.

I would like to return to Winnicott with the suggestion that his notion of "play" is an important contribution to arguments for maintaining openness to the politics of difference, rights and justice. Flax's (1990, 1993) project is to transform Winnicott's notion of transitional spaces into ideas of justice that are not constrained by the imaginary-symbolic as it is translated into spatial effects, and hierarchical and arbitrary valuations of difference. Play is the active explora-tion of individual and social imaginaries, built up in the spaces of everyday life. The challenge around children's rights is to develop legal structures that are robust enough to protect the notion of *thick play*, but flexible enough to guard against its objectification and commodification. Flax (1990, 116) argues that Winnicott's notion of "spaces of play" suggests one of the most important con-tributions to post-Enlightenment thinking because it de-centers reason and logic in favor of "playing with," and "making use of" as the qualities most character-istic of how the self develops. Flax reworks Winnicott's ideas into an approach to justice that jointly applies feminist concerns for the diminution of relations of dominance and post-structural concerns for the play of differences. This is different from postmodern notions of play that sever connections with the past and the future, and other contexts. In such a non-material context, play is trivialized and difference is taken less seriously. This latter conjecture needs some elaboration.

Ideas of difference come together if we are willing to be at *play* – in the classic sense of joining in a dialogue with children and others, but also with ourselves. In this latter condition play is practiced alone, in private, in a space without judge-ment. As part of culture and of young people's lives, play may potentially trans-form the process of social reproduction and, in this sense, play is more than a practice and rehearsal for a known adulthood. The argument that I am making here is that a large part of the potential for young people to transform social reproduction lies with adult propensities to not only leave them the privacy to do so, but also construct consistent guidelines and protection to enable that privacy. The wide-ranging discussions of the last few chapters come together in the thought that there is something about our contemporary cultural moment that perplexes young people's capacity to play. Transitional spaces of unmitigated potential, creativity and imagination are diminishing because they are threatening to adult control and comfort. They disappear with the violent disruptions in various sorts of "war zones" around the world. They are stultified and inappropriately channeled in solitary, efficiently guided electronic games.

Debra Morris (2000, 333) argues that a delicate and even dangerous sort of privacy emerges once transitional spaces are acknowledged because they also establish a right to resist the normalizing powers of science and society:

> For, in delineating a space of nonjudgment, privacy opens up a space of transgression: any reprieve from judgment is a tacit invitation to behave differently, perhaps deviantly. . . . The behavior in question may be "experimental" and daring, or perfectly trivial, but, by necessity, it includes perverse behavior.

Morris (2000) draws on the work of Flax (1993) to acknowledge that justice is a special kind of transitional space that can be exploited in order to cope with the challenge of relating inner and outer reality. She is intent upon establishing an enlightened form of justice that provides a positive political theory of privacy. Her (2000, 324) point is that privacy is judged by social critics as "an incoherent and confused value . . . a perverse and infantile demand, a masculinist prerogative that only enhances the vulnerability and powerlessness of women," while on the other hand, "real privacy . . . has been irreparably damaged: commodified, colonized, dissolved, infiltrated." Morris wants to develop a more complex theory of play and privacy that helps critical thinking about power and powerlessness while at the same time preserving "what is most singular, secret, ineffable, internal, that is, private [about individuals]" (2000, 323).

Play and privacy are intertwined as part of what Winnicott (1965, 179) calls a "protest from within" that issues from the depths of the "non-communicating central self." This has considerable implications for re-visioning children's rights because within the dicta of the CRC, maturity assumes a greater propensity to communicate. For those too young to communicate and those who choose not to, Morris (2000, 346) argues that, in the abstract and with regard to a notion of self whose essential needs cannot be delineated in advance, attitudes "must be those of reverence, silent witness to another's right to privacy." For one thing, such a formulation acknowledges the right to inaction. Doing nothing is a form of action and should not be challenged into a form of indifference.

Privacy, in Morris' formulation, is as a special kind of reprieve from social control, a relief from power, and a removal of a panoptic gaze. To this end, a re-theorizing of Winnicott's object relations helps some feminists and post-structuralists to critique oppressive patriarchal systems that rake through minds, bodies, families and communities. It raises moral issues related to child super-vision and safety, and the control of public and private space. One of Morris' (2000, 330) main points is that private transitional spaces are needed so that certain things may be rendered ambiguous, unseen and unspoken. This does not mean that they cannot be spoken in the sense Nast (2000) articulates about the racist oedipal, but that certain interior/exterior questions are absurd and should not be spoken. Some questions and behaviors around a child's favorite bunny or blanket, for example, are simply forbidden. To present a child with the question ''Did you conceive of this [bunny, blanket] or was it presented to you from without?'' seems absurd and would engender a confused or silent response. Winnicott (1971, 12) points out that it is no less cruel to pose the same kind of question to a teen or adult. To ask a teenager how listening to independent music on a portable player either relates to their sense of self or to that of a larger global experience, for example, is equally absurd, even though they may have a good answer. Morris argues that questions of this kind are ''demoralizing'' because they call into question something that should simply be accepted.

Let me try to elaborate further with a contextual example that relates to tech-nology and new spaces of difference. It is reasonable to assume that if adults use space to contain children's activities and monitor their interactions, resistance may find form in new spaces and communities of creative play. Robins (1996, 94) argues that the new technological environments of virtual reality and the internet may empower young people because they create the illusion that internal and external realities are one and the same in a classic Winnicottian formulation. The problem is that these artificial environments are often created by adults in conformity to normative dictates of pleasure, desire and marketing. But to ask a young person if they are being artificially constructed by this internal environ-ment is ridiculous. Virtual empowerment is none the less a solipsistic affair that encourages a sense of self-containment and self-sufficiency, and involves the denial of the need for external objects. It may be argued further that Disney-like ''imagineers'' produce children's dreams for them. Like the mind-numbing games in *The Discovery Zone* it is possible to become addicted to *Real TV*, LucasArts video games and the internet because they offer the hope of instant gratification (I feel better by watching this spectacle), instant imagination (these graphics are very cool). Clearly, these new technologies and spaces for children offer imaginative possibilities and, as Winnicott put it, an unchallenged potential space that supports infantile illusions of magical creative power. Moreover, in terms of critical moral geographies, some machine interfaces can potentially amplify an amoral indifference to human relationships. What is lost in these new spaces, Robins (1996, 95) argues, is "the continuity of grounded identity that underpins and underwrites moral obligation and commitment." Although they may create places of play and imagination, they do not necessarily recognize the limitation of their control and so the user is less likely to see beyond the imagined world and be able to enter into relationships with actual objects and the material conditions of the world as lived on a day-to-day basis. In many ways, the internet and cyberspaces feed into an addictive propensity to hide in internal spaces while imagining them to be real. But this is only one side of Winnicott's (1988, 107) formulation of transitional space; he also put great emphasis on the moment of disillusionment, which involves "acknowledging a limitation of magical control and acknowledging dependence on the goodwill of people in the external world." Through this process, Robins (1996, 95) argues that a "moral sense" evolves because it "enables the development of capacities for concern, empathy and moral encounter." Because transitional spaces are intermediate spaces they are part of the process through which senses of morality, empathy and concern evolve precisely because they require simultaneous engagement with the real, the external, the off-line.

And so, it is not surprising that in empirical studies of children's use of the computer mediated communications, Holloway and her colleagues find that children's on-line and off-line worlds are mutually constituted (Bingham et al. 1999; Holloway et al. 2000; Valentine and Holloway 2001). The important point that they make, which is at odds with Robins' formulation of cyberspaces' potential for creating amoral landscapes, is that technology does not develop outside of social relations, nor do social relations develop outside of technology. Thus cyberspaces cannot be viewed as invariant and disembodied places, but rather as object relations that materialize for children, like all other object relations, as diverse social practices. Although the emergence of the internet propels moral panics about children's risk of abuse and corruption on-line from strangers or that young people risk addiction to delusional fantasy worlds to the extent that they socially withdraw from the real world, the research of Holloway and her colleagues suggests a much more embedded and embodied interpretation. Anonymity on-line does suggest a certain amount of disembodiedness, but their evidence suggests that the body is still centrally at stake in the production of children's on-line identities. The internet does not have any inherent properties but rather emerges as a different tool for different groups of children in what Holloway and her colleagues call "communities of practice." For some, cyberspace emerges as an important tool for privacy against eavesdropping caregivers, for others it is a tool to develop on-line friendships, for others it is a tool for sociability that enhances and develops everyday off-line social practices, for others it enables the development of hobbies, for still others, it is a tool for playing, having fun or perhaps harassing people (Holloway *et al.* 2001).

Why the issue of the relationship between play, privacy and justice is significant, then, rests on the issue of control and the re-production of a culture of control, scrutiny, zero tolerance, total accountability and instant justice. The control that is exercised over young people's play through, among other things, supervision and the design of play spaces, constrains meanings as well as practices. For Elshtain (1995, 49), in such a society, "the social space for difference, dissent, refusal, and indifference is squeezed out." Thick play is thinned and then squeezed out. This is so because play, at its most radical and important, is as a form of resistance. Resistance is squeezed out. Play is often constructed as irrational and, as such, does not fit well in the rational, instrumental logic that pervades the abstract conceived spaces of today's world. Some behaviors that produce today's spaces for young people are reasoned and increasingly market- and outcome-oriented. Play is not. Transitional spaces are about the kind of imagination and potential that an outcome-oriented world stifles.

If transitional spaces of play, privacy and justice were to reconceptualize the older private–public or newer on-line–off-line dichotomies, then it would be possible to gain a vital source for navigating between the personal and the public, the inside and the outside, the singular and the general. It puts identity politics in a new light. Morris (2000) argues that play, privacy and the public, as conceived by Winnicott, lie simultaneously within the flow of power and they can be used to participate in power. Doing nothing and recalcitrance are not about indifference – although inactions of this kind can be used to hide intolerance, hierarchy and privilege – if they "make thought more penetrating, more sensitive to the complexities of power, powerlessness, and vulnerability" (Morris 2000, 347).

To a large extent, then, I am asking that young people are left alone to discover a morality that is not derived from adult moral panics. I am asking that they are left to themselves, but not in the sense of adult–child separateness, boundedness and individuation that permeates today's culture worlds. Rather, I am asking that young people are let go lightly and given the space and privacy to find themselves with the knowledge that the adult world is very much part of that self. This is to suggest that giving young people space is more than giving them room to play, it is giving them the opportunity for unchallenged and critical reflection on experiences. It is suggesting a space that is more than a "container" of experiences, of mechanistic and hierarchical notions of justice, for those concepts imply that something of unspeakable human importance is lost (Morris 2000, 365).

Connected forms of justice

Winnicott recognizes limitations in the ''self'' proclaiming its freedom from history, geography, technology, meaning and value. In noting this, James Glass (1995, 174) points out that his argument has stronger claims, ones Winnicott does not make, which focus not only on an appreciation of otherness and tolerance but also on new forms of justice. It may be argued, then, that a neo-Winnicottian formulation of the self embraces, to take one example, the spatial and unjust effects of the political unconscious so eloquently articulated by Nast (2000) as a hidden racist oedipal construct. Such ownership directs us to understand more fully the differences within and without.

If we agree with Wilson (1980) that an ideological propensity toward reason undergirded the construction of childhood beginning in the early industrial period, then we need to also recognize how this same ideology constitutes issues of justice. Flax (1993) suspects that claims to an impartial, definitive and mechanistic version of justice based on logic and reason are also based upon claims to liberal autonomy, individual control and, I would add, exclusion and the separation of children from adults. Justice constituted in these terms may serve specific and articulated rights but not necessarily people's everyday lives and welfare. A discourse of negotiation open to all is a model of justice that seeks neither to marginalize nor prioritize any one point of view: neither that of children nor that of adults. Flax (1993, 340) argues that because transitional spaces serve as defenses against the fear of multiplicity, ambivalence and uncertainty that leads to discourses that collapse all worlds into one, then they also support and sustain ethical relations, defusing the paranoia and moral panics that corrupt rational thought. Traditional versions of justice typically involve either some hierarchical and arbitrary valuation of difference or, less often, some uniform treatment of difference, that while appearing more equitable, disguise the real and ongoing forms of domination that exist in the construction of gender, ethnicity, and class (cf. Young 1990b). Rather than being a unitary concept grounded in some external truth, Flax (1993, 112) views justice as a process made up of interrelated practices: ''Differences must somehow be confronted, accommodated, or harmonized within a whole that strives to achieve the good(s) for all and in which relations of dominance are minimized.'' In order for this process not to result in asymmetric dualisms, and for difference not to be used as the justification for hierarchies, a mechanism for consensus building must be available. At one level, transitional spaces fulfill this function while at the same time allowing differences to continue to have value within an individual's existence.

Diduck (1999, 121) points out that a re-conception of (legal) justice that recognizes dependence (that is, the conditions and relationships of connection) may violate the liberal principles of autonomy, equality and universality upon which Western law is based. For children to get rights and justice that are appropriately constituted, then we must depart from singular notions that embrace only an autonomous, independent subject that is unchildlike. Diduck argues that if this is the case, then the autonomous self needs to be reconstituted in terms of relationships rather than disconnections:

> This presentation recognizes that a legal subject is never abstracted and individuated, but rather always exists as part of his or her context and relationships – even relationships based upon love rather than upon rights or exchange . . . [and] also attributes an agency to the subject to play a part in his or her subjectification. . . . Childhood, like adult-hood, cannot be universalized in this view, and we must speak of child-hood and adulthood in the plural.
>
> (Diduck 1999, 121)

This does not negate Morris' concerns for new theories of privacy and the need for a solitude that protects from the intrusion of adults or the intervention of the state, but simultaneously broaches the notion that child–adult boundaries are permeable and infused with meaning.

In play there are no boundaries between subject and object because play trans-forms subject and object into and through process. Understood as process, justice is one way for individuals to manage the strain of being simultaneously public and private, alone and related to others (Flax 1993, 123). Engaging in just practices ("doing justice" as Diduck calls it) in this way offers the possibility of modes of relatedness with others that challenge notions of autonomy which exclude dependence and notions of welfare that exclude individuality. But, as Diduck (1999, 121) points out, justice on these terms may cause problems for able-bodied white adult males whose subjectivity rests in a liberal autonomous tradi-tion based upon mechanistic and static notions of reason and logic. For others, and particularly children, she suggests that there are fewer problems with a notion of a justice that is playful because it embraces dialogue, dependence and welfare as well as independence, privacy, individualistic autonomy and reason.

A museum of childhood

On Edinburgh's High Street, a narrow three-story building houses Scotland's Museum of Childhood. It is advertised as "a favourite with adults and children alike – it is a treasure house, crammed full of objects telling of childhood, past and present." The museum is a repository for all sorts of childhood paraphernalia and memorabilia. Some seven years ago, after discovering the location of this venerable institution while looking for something else in the *Yellow Pages*, I set aside a whole afternoon for exploration and discovery. In the museum, past childhoods are celebrated from behind dioramas and glass display cases containing rather ugly and decrepit toys. Some of the show cases highlight Victorian ingenuity in the making of mechanical toys or the intricacy of doll's houses, but none celebrate children's play. Nor do any articulate the relations between children's work, play, education and discipline. Jo Boyden (1990, 185) articulates similar feelings on a visit to London's Museum of Childhood. These are very depressing and quite boring entombments of adult nostalgia and past commodifications of childhood. I was glad to be away from Edinburgh's museum after less than an hour.

Why do museums of childhood not celebrate play? In the simulated cave of Canada's first children's museum, we tried something different. We tried engaging children at a different level, one that didn't worry about authenticity or protocol. We wanted them to have fun and to perhaps learn something along the way. I do not know if any of my young playmates learnt anything from Stuart-the-caveman, but he learnt something from them. It seems to me that as researchers studying children, we are the ones who are often looked at, gazed at and inspected. This gaze, which is neither a threat nor a retaliation, makes us conscious of ourselves, leading to our need to turn this gaze around and look at ourselves. What is the moral integrity here? Who was Stuart-the-caveman? Well, some of my fascination with young people may be simply a device to hold on to a seemingly unchanging certainty – such as the innocence of childhood – somewhere outside of my own ''fake'' experiences. For me, I believe the immediacy of fieldwork with children requires that I divest myself of preconceived and problematic conceptions such as childhood, and attend to the needs of potential new friends. After years of participant observations, ethnographies, oral histories and all the fancy names I use for being with young people, I know that in the immediacy of the moment all I want is to be liked, to be a friend. The theorizing and answering of sticky research questions comes later if at all. Canada's first children's museum touted itself as being child-friendly. That is the key. Stuart-the-caveman was available for play and that is all that the immediacy of the moment required. I plan on taking my kids back to that museum when next we visit Ontario. We'll be going to Scotland in the not too distant future also. I think we'll avoid the Museum of Childhood on Edinburgh's High Street.

In a previous book, I used play as a metaphor for engaging in dialogue with others, our environments, ourselves and our systems of justice. I still believe that reconceptualizing the relationship between play and justice is important so I do not want to diminish the importance of dialogue as a metaphor for play, and play as a metaphor for important aspects of lived experience. Rather,

I want to project insight from that onto concerns for young people that are at once more global and more tied to place, that accommodate difference, and that prescribe new kinds of justice, democracy and public spheres that do not exclude children or attempt to survey their private worlds.

Throughout the process of writing and revising this book, I experienced doubts and reservations about its relevance and value to young people. One reviewer of an early manuscript lamented the lack of children's voices throughout the text. I refrained from including children's voices in the revisions because I am concerned about tokenism, but I am also concerned about surveying and appropriating children's private worlds. This book is about me and the adult world from which I derive my sensibilities and pretensions. I am convinced that our imperious judgments as to what matters in being an adult help explain the current crises of childhood. Young people's expressions of self are often sandwiched and squeezed into something that is palatable to adult sensibilities. As I watched and listened to the four children in Mexico City's Zócalo those many years ago I sensed the irony of their play as I saw it, sandwiched between the Spanish cathedral and the pyramid of Tenochtitlán. The former a monolithic icon of imperialism and the latter an attempt to represent an equally repressive culture and tie it into new circuits of capital through tourist dollars. I argue throughout the pages of this book that the cultural and economic politics of global oppression require an insidious model for the tunes that children play so that they are soothing to adult ears. Attempts to buttress romantic perceptions of children's innocence, naïveté and lack of discipline are really impositions of adult, desiccated ways of knowing. Herein lies our audacity and the hypocritical and potentially destructive basis of adult domination and the faceless processes of globalization. As these thoughts go through my mind years later when I reflect on the children of the Zócalo, I remember that the baby's bowl remained for the most part empty but the children played on regardless. To this day I am shamed by my inability to break through the space that their cacophony created, to walk into their circle and deposit a few pesos in the bowl. At some point I need to become vulnerable, I need to stop ruminating and theorizing. I need to reach out from behind my camera and notebook to the basic humanity at the heart of every encounter. I need to embrace that humanity in the same way those young children in the simulated cave reached out and playfully embraced Stuart-the-caveman. But in the face of these young people's humanity a horror looms that won't recede and so comes forth late at night: how do I situate myself? When I open my heart to a listener or pour forth in a confessional essay, recounting chasms that cut deep into the landscapes of my being, do I, the participant who is much more comfortable as an observer, still remain behind the lens of my camera, buried underneath simulated furs?

Note

1 To date, the United States and Somalia are the only member countries of the UN not to ratify the CRC and make this legal commitment to children. The United States has signalled its intention to ratify by formally signing the Convention but requires an extensive examination and scrutiny of treaties before proceeding to ratify. According to the UNICEF website, this examination, which includes an evaluation of the degree of compliance with existing law and practice in the country at state and federal levels, can take several years – or even longer if the treaty is portrayed as being controversial or if the process is politicized (http://www.unicef.org/crc/faq.htm#009). Marilyn Ivy (1995, 102) argues that this politicization is related to contention between the rights of children on the one hand and transnationalization of capitalism on the other. The implication is that the profits of transnational corporations will be impacted by strong international laws barring exploitation of young people's labor.

EPILOGUE

At 9.20 am on March 5, 2001, as I fastidiously engaged the final corrections for this manuscript, another moral panic erupted in the United States. The epicenter was only a few miles from where I was working, but the panic rapidly poured out through national and global news media like a super-heated gas cloud. A 15-year-old, Charles Andrew Williams, opened fire on the campus of Santana, a suburban San Diego high school, killing two students and injuring thirteen other people. For eight minutes, Williams terrorized the school, firing randomly inside a bathroom and around a campus courtyard. Reports suggested he calmly reloaded his eight-shot pistol with a wistful smile on his face, discharging at least thirty shots into scampering groups of students, teachers, and staff members. The incident is the nation's worst since Columbine, with a context so similar that it warrants some comments on what seems to be a normalizing and, I argue, mythic geography of fear. Headings in the *Los Angeles Times* the day after the shooting reported that the ''Shooting Suspect's Heart was Still in His Small Maryland Hometown.'' And on the next day, another headline proclaimed ''Nightmare Evolves From the Suburban Dream'' (*Los Angeles Times*, March 8, front page, column 1). A specific geography is evoked by these headlines, but conflated with them are speculations on the nature of Santee, the community of Santana high school, that summon fears of the unraveling of the American dream around its young people and where they go to school.

As the week proceeded to the weekend, and reporters, politicians, parents and kids scrambled to make sense of the tragedy, fears of copycat violence further exacerbated the moral panic. School officials across the nation locked down their campuses, and a police presence was visible in every San Diego high school and some elementary schools. Because Williams allegedly told friends and at least one adult who disregarded him that he intended to shoot up his campus, any kind of threat or joke was taken very seriously. Young people across the nation were suspended, sent for counseling and even jailed amid calls for policies to protect ''snitchers'' from litigation. Who are these potential perpetrators of campus violence? Despite the complete inability of psychometric models focused on developmental deviance to predict violent behavior, the press lauds long-term quantitative studies that attempt to identify "risk factors" in young people. According to the most recent report on youth violence from the office of United States Surgeon General, the greatest early risk factors are unspecified "general delinquency" and "substance abuse" followed closely by simply "being male." Are all young men under suspicion? It would seem so.

What I want to point out here, and what came across as part of the barrage of media reporting on the Santana tragedy, is the white suburban contexts within which young violent men live. I do so to underscore the focus in this book on morally contested geographies that hide certain people and places while mytholo-gizing others. On the night of the Santana shooting, the Mayor of Santee, in words that echoed some post-Columbine sentiments, suggested that "If it can happen here it can happen anywhere." That "anywhere" is the mythic place of white middle-class America – suburbs, small towns, edge communities, country villages. "Living in Santee," noted a parent in a local news report, "is like living in the '50s." It is the place of the mythic American nuclear family.

In the "Nightmare Evolves From the Suburban Dream" article, *LA Times* reporters Scott Gold, Elaine Gale and Richard Marosi argue that "by income, by location, ethnicity or almost any measure you could apply, Andy Williams' hometown lies right in the crosshairs." Santee is a classic dormitory community suburb of San Diego that grew rapidly in the last decade as people sought the American dream with relatively affordable modest tract homes. The community has little or no economic base, with most people commuting to downtown San Diego or the Marine Corps Air Station at Miramar. The *LA Times* reporters go on to argue (March 8, 2001, A–14) that most of the notorious places where school killings occur – Moses, Lake, Fort Gibson, Pearl, West Paducah, Joneboro, Springfield, Conyers, Littleton – are unheard-of small communities prior to the incident. "All a town really needs to be a potential target," they argue, "is a population of teenage boys and guns."

There is a disturbingly sexist, racist and adultist rhetoric to these fears. Authorities, policy-makers and reporters who portrary school shootings as caused by "lone teens," "delinquent youth," "angry young men" and other demeaning stereotypes, miss the point that the moral panic is not about the shoot-ings themselves. After all, according to a recent *New York Times* review, a random public shooting occurs in the United States every ten days but few warrant the attention garnered by Williams in Santee, or Harris and Klebold in Littleton. Perpetrators of "everyday" shootings – usually middle-aged, affluent, white and disgruntled men – make the local news but are soon forgotten. African-American, Asian and Latino youth violence is most often gang related or caused by personal disputes at schools, but these – mostly inner-city – minority crimes also rarely warrant front page news. Rather, moral panics ensue when the killings are of white students, by white students, who are ostensibly contextualized by the American small town/suburban dream. Are we not concerned with this kind of reporting as a process of racialization, where race is constructed as a means of differentiating and valuing "white" people above those of color? Importantly, there is a conflation of place (suburban schoolyard) with race (white).

In a weekend opinion piece in the *LA Times*, Mike Males (March 11, 2001, M6) points out that schoolyard killings are much less common when compared to the aggregate statistics for the United States: "Of the 150,000 Americans murdered by gunfire in the last decade, perhaps 150 were killed in or around a school, and only a fraction were white youths. If the US's overall murder rate was as low as that in high schools, America would be safer than Sweden." The moral panic is not because these kinds of shootings are unprecedented. Males goes on to enumerate forgotten school shootings going back to the 1970s. Although not rare when compared to national statistics, white-student shootings in suburban schools tear apart the fabric of the American myth, portending the imminent rending of society's hallowed places and its future in its young people who live there. By focusing on the need to uphold this myth that never was, journalists, policy-makers and other government officials highlight the need for more studies to discover "risk factors" and the universal causes of teen violence while conveniently forgetting larger social and spatial complexities. A scientific study to identify why teens become deviant may be fraught with all the developmental and naturalizing assumptions outlined throughout this book, but it is relatively simple and mechanistic, requiring neatly set up hypotheses, sampling frames to identify appropriate populations, and robust testing procedures. The larger social and spatial contexts of young people's lives – poverty, sexism, racism, adult violence, exclusion – are conveniently hidden behind august proclamations by scientists needing more data, more subjects, and more funds for study.

In the meantime, President George W. Bush cites the role of the family in stemming school violence. In his first reaction to the shooting, Bush argued that private actions and responsibilities are far more important than public policies in preventing such tragedies. After the Columbine shooting he averred "Of course there is going to be reactions – pass a law. The big law is the universal law – how do mothers and dads do their jobs" (*The Denver Post*, April 22, 1999, Thursday 2D Edition). By taking this position, Bush suggests that his new government, like that of his father, will rest culpability squarely on the shoulders of those who can least afford to bear it: impoverished and beleaguered families. Charles "Jeff" Williams – Andrew's father – is a single parent who had recently moved to Santee in search of a better life. Although his relationship with Andrew has yet to be thoroughly dissected, first reports suggest a healthy and strong sense of family. Andrew was an honors student who seemed happy although his recent move from Maryland was of some concern. There is some media speculation, none the less, that single parenting in general, and single fathering in particular, is deviant. The universal law, it seems, is not only that mothers and fathers do their jobs but that they uphold the institution of the nuclear family.

The sense of outrage, the moral panic, evolves in part from the media spectacle of school shootings in seemingly tight-knit, family-oriented communities. How can something so horrifying happen to a community that so resembles middle America? How can the perpetrator be so like my son? Such perspectives are disturbing because they erase and homogenize the myriad social and spatial contexts that make up the lives of young people and the complexities of their families and communities. The outcome of these lives can no less be predicted than can the identity of a potential shooter. The complexity of these lives varies across places according to processes accruing to labor market potentials, firearm ownership rates, real estate practices, migration, school board policies, police surveillance tactics, built environments and a host of other salient contexts. The case of Andrew Williams, his family, his friends and his town reminds us that the entire United States is deeply embroiled in contesting the spaces of youth identity, even as ''whiteness'' and ''youth violence'' together serve as a lightning rod for moral panics.

Stuart C. Aitken
San Diego, Wednesday, March 14, 2001

BIBLIOGRAPHY

Adams, Paul, Leila Berg, Nan Berger, Michael Duane, A.S. Neill and Robert Ollendorff. 1971. *Children's Rights: Towards the Liberation of the Child*. New York: Praeger Publishers.

Ainsworth, Mary D.S. 1972. Variables influencing the development of attachment. In C.S. Lavatelli and F. Stender (eds) *Readings in Child Behavior and Development*. New York: Harcourt Brace Jovanovich.

Aitken, Stuart C. 1994. *Putting Children in Their Place*. Association of American Geographers, Washington, DC: Edwards Bros.

Aitken, Stuart C. 1998. *Family Fantasies and Community Space*. New Brunswick, NJ: Rutgers University Press.

Aitken, Stuart C. 1999. Scaling the light fantastic: geographies of scale and the web. *Journal of Geography* 98: 118–27.

Aitken, Stuart C. 2000a. Play, rights and borders: gender bound parents and the social construction of children. In Sarah Holloway and Gill Valentine (eds) *Children's Geographies: Playing, Living and Learning*, pp. 119–38. London and New York: Routledge.

Aitken, Stuart C. 2000b. Fear, loathing and space for children. In John R. Gold and George Revill (eds) *Landscapes of Defense*, pp. 48–67. Oxford: Oxford Brookes University Press.

Aitken, Stuart C. 2001. Playing with children: social reproduction and the immediacy of fieldwork. *Geographical Review*. Forthcoming.

Aitken, Stuart C. and Thomas Herman. 1997. Gender, power and crib geography: from transitional spaces to potential places. *Gender, Place and Culture: A Journal of Feminist Geography* 4(1): 63–88.

Aitken, Stuart C. and Christopher Lee Lukinbeal. 1997. Mobility, road geographies and the quagmire of terra infirma. In Steven Cohen and Ina Rae Hark (eds) *Road Movies*, ch. 16, pp. 349–70. London: Routledge.

Aitken, Stuart C. and Christopher Lee Lukinbeal. 1998. Of heroes, fools and fisher kings: cinematic representations of street myths and hysterical males. In Nicholas R. Fyfe (ed.) *Images of the Street*. London and New York: Routledge.

190

Aitken, Stuart C. and Leo E. Zonn. 1993. Weir(d) sex: representations of gender–environment relations in Peter Weir's *Picnic at Hanging Rock* and *Gallipoli*. *Environment and Planning D: Society and Space* 11: 191–212.

Anderson, Kay. 1997. A walk on the wild side: a critical geography of domestication. *Progress in Human Geography* 21(4): 463–85.

Anzaldúa, Gloria. 1987. *Borderlands/La Frontera: The New Mestiza.* San Francisco: Aunt Lute Books.

Appleton, Jay. 1975. *The Experience of Landscape.* London: John Wiley.

Archard, David. 1993. *Children: Rights and Childhood.* New York and London: Routledge.

Ariès, Philippe. 1962. *Centuries of Childhood: A Social History of Family Life.* New York: Alfred A. Knopf.

Armstrong, David. 1983. *Political Anatomy of the Body: Medical Knowledge in Britain in the Twentieth Century.* Cambridge: Cambridge University Press.

Armstrong, David. 1986. The invention of infant mortality. *Sociology of Health and Illness* 8: 211–32.

Bakan, David. 1990, previously published in 1958. *Sigmund Freud and the Jewish Mystical Tradition.* New York: Free Association Books.

Barnes, Trevor and James Duncan (eds). 1992. *Writing Worlds: Discourse, Text and Metaphor.* London: Routledge.

Beazley, Harriott. 2000. Home sweet home?: street children's sites of belonging. In Sarah Holloway and Gill Valentine (eds) *Children's Geographies: Playing, Living and Learning*, pp. 194–212. London and New York: Routledge.

Beazley, Harriott. 2001. Street kids, identity, resistance and public space in Yoyakarta, Indonesia. *Antipode.* In review.

Bell, David and Gill Valentine (eds). 1995. *Mapping Desire.* New York and London: Routledge.

Bingham, Nick, Sarah L. Holloway and Gill Valentine. 1999. Where do you want to go tomorrow? Connecting children to the internet. *Society and Space* 17: 655–72.

Blades, Mark, J.M. Blaut, Zhra Darvizeh, Silvia Elguea, Steve Sowden, Bhiru Soni, Christopher Spencer, David Stae, Roy Surajpaul and David Uttal. 1998. A cross-cultural study of young children's mapping abilities. *Transactions of the Institute of British Geographers* 23(2): 269–77.

Blaut, Jim. 1971. Space, structure and maps. *Tidjschrift Voor Economische En Sociale Geografie* 62: 1–4.

Blaut, Jim. 1991. Natural mapping. *Transactions, Institute of British Geographers, New Series* 16: 55–74.

Blaut, Jim. 1997. Piagetian pessimism and the mapping abilities of young children: a rejoinder to Liben and Downs. *Annals of the Association of American Geographers* 87(1): 168–77.

Blaut, Jim. 1999. Maps and spaces. *The Professional Geographer* 51(4): 510–15.

Blaut, Jim and David Stea. 1971. Studies of geographic learning. *Annals of the Association of American Geographers* 61: 387–93.

Blaut, Jim and David Stea. 1974. Mapping at the age of three. *Journal of Geography* 73: 5–9.

Blaut, Jim, G. McCleary and A. Blaut. 1970. Environmental mapping in young children. *Environment and Behavior* 2(3): 335–49.

Bloch, J. 1974. Rousseau's reputation as an authority on childcare and physical education in France before the Revolution. *Paedagogica Historica* 14: 5–33.

Bluebond-Langer, M. 1978. *The Private Worlds of Dying Children.* Princeton, NJ: Princeton University Press.

Bluebond-Langer, M. D. Perkel and T. Goertzel. 1991. Paediatric cancer patients' peer relationships; the impact of an oncology camp experience. *Journal of Psychosocial Oncology* 9(2): 67–80.

Blum, Virginia and Heidi Nast. 1996. Where's the difference? The heterosexualization of alterity in Henri Lefebvre and Jacques Lacan. *Environment and Planning D: Society and Space* 14: 559–80.

Bondi, Liz. 1996. In whose words? On gender identities, knowledge and writing practices. *Transactions of the British Institute of Geographers* 22(2): 245–58.

Bondi, Liz. 1999. Stages on journeys: some remarks about human geography and psychotherapeutic practice. *The Professional Geographer* 51(1): 11–24.

Bourdieu, Pierre. 1984. *Distinction: A Social Critique of the Judgement of Taste.* London and New York: Routledge.

Bowlby, John. 1951. *Maternal Care and Mental Health.* Geneva: World Health Organization.

Bowlby, John. 1988. *A Secure Base.* New York: Basic Books.

Bowlby, Sophie, Sally Lloyd Evans and Robina Mohammad. 1998. The workplace: becoming a paid worker: images and identity. In Tracey Skelton and Gill Valentine (eds) *Cool Places: Geographies of Youth Cultures*, pp. 229–48. London and New York: Routledge.

Boyd, William. 1962. *The* Emile *of Jean-Jacques Rousseau.* New York: Columbia University, Bureau of Publications.

Boyden, Jo. 1990. Childhood and policy makers: a comparative perspective on the globalization of childhood. In Allison James and Allan Prout (eds) *Constructing and Reconstructing Childhood,* pp. 184–215. London and New York: Falmer.

Bradshaw, John. 1988. *Bradshaw On: The Family.* Florida: Health Communications Incorporated.

Brainerd, C.J. 1978. The stage question in cognitive-development theory. *The Behavioral and Brain Sciences* 2: 173–213.

Brandtstädter, Jochen. 1990. Development as a personal and cultural construction. In G.R. Semin and K.J. Gergen (eds) *Everyday Understanding: Social and Scientific Understanding*, pp. 83–107. London: Sage Publications.

Breitbart, Myrna M. 1998. ''Dana's mystical tunnel'': young people's designs for survival and change in the city. In Tracey Skelton and Gill Valentine (eds) *Cool Places: Geographies of Youth Cultures*, pp. 305–27. London and New York: Routledge.

Buchholz, Ester Schaler. 1997. *The Call of Solitude: Alonement in a World of Attachment.* New York: Simon & Schuster.

Buckingham, David. 1994. Television and the definition of childhood. In Brian Mayall (ed.) *Children's Childhoods: Observed and Experienced.* London: Falmer.

Bunge, William W. 1971. *Fitzgerald: The Geography of Revolution*. Cambridge, MA: Schenkman.

Bunge, William W. 1973. The geography. *The Professional Geographer* 25(4): 331–7.

Bunge, William W. 1977. The point of reproduction: a second front. *Antipode* 9: 60–76.

Bunge, William W. 1979. Perspective on theoretical geography. *Annals of the Association of American Geographers* 69: 128–32.

Bunge, William W. and R. Bordessa. 1975. *The Canadian Alternative: Survival, Expeditions and Urban Change*. Geographical Monographs, No. 2. Toronto: York University.

Bunting, Trudi. 1983. Urbanism in children: how school-age children evaluate urban environments. *Ontario Geography* 21: 3–28.

Bunting, Trudi. 1986. Gender-related differences in the development of environmental dispositions. *Ohio Geographers: Recent Research Themes* 14: 89–108.

Bunting, Trudi and L.R. Cousins. 1983. Environmental personality in school-age children: development and application of the children's environmental response inventory. *Journal of Environmental Education* 15: 3–10.

Bunting, Trudi and L.R. Cousins. 1985. Environmental dispositions among school-age children: a preliminary investigation. *Environment and Behavior* 17(6): 725–68.

Bunting, Trudi and T. Semple. 1983. The development of an environmental response inventory for children. In A.D. Seidel and S. Danford (eds) *Environmental Design, Research Theory and Application*, pp. 273–83. Washington, DC: Environmental Design Research Association.

Burman, Erica. 1994. *Deconstructing Developmental Psychology*. London and New York: Routledge.

Butler, Judith. 1990. *Gender Trouble: Feminism and the Subversion of Identity*. New York: Routledge.

Butler, Judith. 1993. *Bodies that Matter: On the Discursive Limits of "Sex"*. New York: Routledge.

Butler, Ruth. 1999. The body. In Paul Cloke, Philip Crang and Mark Goodwin (eds) *Introducing Human Geographies*, pp. 238–46. London: Arnold.

Callard, Felicity. 1998. The body in theory. *Society and Space* 16: 387–400.

Cantwell, N. 1989. A tool for implementation of the UN convention. In Radda Barnen (ed.) *Making Reality of Children's Rights*, pp. 36–41. International Conference on the Rights of the Child.

Chomsky, Noam. 1965. *Aspects of the Theory of Syntax*. Cambridge, MA: MIT Press.

Cisneros, Sandra. 1989. *The House on Mango Street*. New York: Random House.

Clifford, James. 1988. *The Predicament of Culture: Twentieth Century Ethnography, Literature and Art*. Cambridge, MA: Harvard University Press.

Coleman, James. 1990. Social institutions and social theory. *American Sociological Review* 55: 333–9.

Coontz, Stephanie. 1992. *The Way We Never Were: American Families and the Nostalgia Trap*. New York: Basic Books.

Cunningham, Hugh. 1995. *Children in Western Society Since 1500*. London: Longman.

Darwin, Charles. 1887. A biographical sketch of an infant. *Mind* 7.

Davis, Karen. 1988. What is ecofeminism. *Women and Environments* 10: 4–6.

Davis, M. and D. Wallbridge. 1981. *Boundary and Space: An Introduction to the Work of D.W. Winnicott*. New York: Brunner/Mazel Publishers.

Davis, Susan G. 1997. Space jam: family values in the entertainment city. Paper presented at the American Studies Annual Meeting, Washington, DC.

Demos, John. 1970. *A Little Commonwealth: Family Life and the Life Course*. New York: Oxford University Press.

Diduck, Alison. 1999. Justice and childhood: reflections on refashioning boundaries. In M. King (ed.) *Moral Agendas for Children's Welfare*. London and New York: Routledge.

Donaldson, M. 1978. *Children's Minds*. London: Fontana.

Downs, Roger G. 1985. The representation of space: its development in children and in cartography. In R. Cohen (ed.) *The Development of Spatial Cognition*. Hillsdale, NJ: Lawrence Erlbaum.

Downs, Roger G. and Lynn S. Liben. 1987. Children's understanding of maps. In P. Ellen and C. Thinus-Blanc (eds) *Cognitive Processes and Spatial Orientation in Animal and Man: Volume 2, Neurophysiology and Developmental Aspects*, pp. 202–19. Dordrecht, the Netherlands: Martinus Nijhoff.

Downs, Roger G. and Lynn S. Liben. 1991. The development of expertise in geography: a cognitive-development approach to geographic education. *Annals of the Association of American Geographers* 81: 304–27.

Downs, Roger G. and Lynn S. Liben. 1993. Mediating the environment: communicating, appropriating and developing graphic representations of place. In R.H. Wozniak and K. Fischer (eds) *Development in Context: Acting and Thinking in Specific Environments*, pp. 155–81. Hillsdale, NJ: Lawrence Erlbaum Associates.

Downs, Roger G. and Lynn S. Liben. 1997. The final summation: the defense rests. *Annals of the Association of American Geographers* 87(1): 178–80.

Downs, Roger G., Lynn S. Liben and D.G. Daggs. 1988. On education and geographers: the role of cognitive developmental theory in geographic education. *Annals of the Association of American Geographers* 78(4): 680–700.

Downs, Roger G., Lynn S. Liben and D.G. Daggs. 1990. Surveying the landscape of developmental geography: a dialogue with Howard Gardner. *Annals of the Association of American Geographers* 80(1): 124–8.

Duncan, James and David Ley. 1993. Introduction: representing the place of culture. In James Duncan and David Ley (eds) *Place/Culture/Representation*, pp. 1–21. London: Routledge.

Dyck, Isabel. 1996. Mother or worker? Women's support networks, local knowledge and informal child care strategies. In Kim England (ed.) *Who Will Mind the Baby? Geographies of Child Care and Working Mothers*, pp. 123–85. London and New York: Routledge.

Elshtain, Jean. B. 1990. The family in political thought: democratic politics and the question of authority. In J. Sprey (ed.) *Fashioning Family Theory*, pp. 51–66. Newbury Park, CA: Sage Publications.

Elshtain, Jean. B. 1995. *Democracy on Trial*. New York: Basic Books.

England, Kim (ed.). 1996. *Who Will Mind the Baby? Geographies of Child Care and Working Mothers*. London and New York: Routledge.

Entrikin, Nicholas. 1994. Moral geographies: the planner in space. *Geographical Research Forum* 14: 113–19.

Erikson, Erik H. 1969. *Childhood and Society*. New York: Norton.

Erikson, Erik H. 1977. *Toys and Reasons*. New York: Norton.

Everhart, Robert. 1983. *Reading, Writing and Resistance: Adolescence and Labor in a Junior High School*. Boston: Routledge.

Fernández Kelly, Patricia. 1994. Towanda's triumph: social and cultural capital in the transition to adulthood in the urban ghetto. *International Journal of Urban and Regional Research* 18: 88–111.

Fielding, Shaun. 2000. Walking on the left!: children's geographies and the primary school. In Sarah Holloway and Gill Valentine (eds) *Children's Geographies: Playing, Living and Learning*, pp. 230–44. London and New York: Routledge.

Flax, Jane. 1990. *Thinking Fragments: Psychoanalysis, Feminism, and Postmodernism in the Contemporary West*. Berkeley: University of California Press.

Flax, Jane. 1993. *Disputed Subjects: Essays on Psychoanalysis, Politics, and Philosophy*. New York and London: Routledge.

Foucault, Michel. 1970. *The Order of Things: An Archaeology of the Human Sciences*. New York: Vintage Books.

Foucault, Michel. 1977. *Discipline and Punish: The Birth of the Prison*. New York: Vintage Books.

Foucault, Michel. 1980. *The History of Sexuality, Volume 1: An Introduction*. Trans. Robert Hurley. New York: Vintage/Random House.

Fraser, Nancy. 1989. *Unruly Practices: Power, Discourse and Gender in Contemporary Social Theory*. Minneapolis: University of Minnesota Press.

Freud, Sigmund. 1961a. *Civilization and Its Discontents*. New York: Norton.

Freud, Sigmund. 1961b. *The Ego and The Id*. In James Strachey, translator and editor, *Standard Edition*, 19, pp. 3–66. London: Hogarth.

Freud, Sigmund. 1965. *New Introductory Lectures on Psychoanalysis*. Trans. J. Strachey. New York: Norton.

Friedberg, Anne. 1993. *Window Shopping: Cinema and the Postmodern*. Berkeley: University of California Press.

Frønes, Ivar. 1994. Dimensions of childhood. In Jens Qvortrup, Margarita Bardy, Giovanna Sgritta and Helmut Wintersberger (eds) *Childhood Matters: Social Theory, Practice and Politics*, pp. 145–64. Aldershot, UK: Avebury Press.

Fyfe, A. 1993. *Child Labor: A Guide to Project Design*. Geneva: International Labor Office.

Fyfe, Nicholas (ed.). 1998. *Images of the Street: Planning, Identity and Control in Public Places*. London and New York: Routledge.

Gagen, Elizabeth. 2000a. An example to us all: child development and identity construction in early 20th century playgrounds. *Environment and Planning A* 32(4): 599–616.

Gagen, Elizabeth. 2000b. Playing the part: performing gender in America's playgrounds. In Sarah Holloway and Gill Valentine (eds) *Children's Geographies: Playing, Living and Learning*, pp. 213–29. London and New York: Routledge.

Gergen, K.J., G. Gloger-Tippelt and P. Berkowitz. 1990. The cultural construction of the developing child. In G.R. Semin and K.J. Gergen (eds) *Everyday Understanding: Social and Scientific Understanding*, pp. 108–29. London: Sage Publications.

Glass, James. 1995. *Psychosis and Power: Threats to Democracy in the Self and the Group.* Ithaca, NY, and London: Cornell University Press.

Goethe, Johann Wolfgang von. 1977, originally published in 1796. *William Meister's Years of Apprenticeship.* Trans. H.M. Waidson, 6 vols. Zurich: Calder.

Goldman, Dodi. 1993. Introduction. In Dodi Goldman (ed.) *In Ones Bones: The Clinical Genius of Winnicott.* London: Jason Aronson, Inc.

Gollaher, David L. 2000. *Circumcision: A History of the World's Most Famous Surgery.* New York: Basic Books.

Golledge, Reginald G. 1978. Learning about urban environments. In T. Carlstein, D. Parkes and Nigel Thrift (eds) *Timing Space and Spacing Time I: Making Sense of Time*, pp. 76–98. London: Edward Arnold.

Golledge, Reginald G. and Robert J. Stimpson. 1997. *Spatial Behavior: A Geographic Perspective.* New York: Guilford Press.

Golledge, Reginald G., T.R. Smith, J.W. Pellegrino, S. Doherty and S.P. Marshall. 1985. A conceptual and empirical analysis of children's acquisition of spatial knowledge. *Journal of Environmental Psychology* 5: 125–52.

Golledge, Reginald G., A.J. Ruggles, James W. Pellegrino and Nathan Gale. 1993. Integrating route knowledge in an unfamiliar neighborhood. *Journal of Environmental Psychology* 13: 293–307.

Golledge, Reginald G., Valerie Dougherty and Scott Bell. 1995. Acquiring spatial knowledge: survey versus route-based knowledge in unfamiliar environments. *Annals of the Association of American Geographers* 85(1): 134–58.

Gomes, P.G. and C.A. Mabry. 1991. Negotiating the world: the developmental journey of African-American children. In J. Everett, S.S. Chipungu and B.R. Leashore (eds) *Child Welfare: An Africentric Perspective*, pp. 156–86. New Brunswick, NJ: Rutgers University Press.

Goussault. 1693. *Le Portrait d'une Honnête Femme.*

Gregory, Derek. 1994. *Geographical Imaginations.* Cambridge, MA: Blackwell.

Gregson, Nicky and Michelle Lowe. 1995. Home-making: on the spatiality of daily social reproduction in contemporary middle-class Britain. *Transactions of the Institute of British Geographers* 20: 224–35.

Grieshaber, Susan. 1998. Constructing the gendered infant. In Nicola Yelland (ed.) *Gender in Early Childhood*, pp. 15–35. London and New York: Routledge.

Griffin, Christine. 1985. *Typical Girls?* London and Boston: Routledge & Kegan Paul.

Griffin, Christine. 1993. *Representations of Youth: The Study of Youth and Adolescence in Britain and America.* Cambridge: Polity Press.

Grossman, David. 1998. Trained to kill. *Christianity Today* 42(9): 30–41.

Grosz, Elisabeth. 1990. *Jacques Lacan: A Feminist Introduction.* London and New York: Routledge.

Grosz, Elisabeth. 1994. *Volatile Bodies: Toward a Corporeal Feminism.* Bloomington: Indiana University Press.

Grosz, Elisabeth. 1995. *Space, Time and Perversion*. London and New York: Routledge.

Habermas, Jurgen. 1984. *The Theory of Communicative Action: Vol. 1*. Boston: Beacon Press.

Habermas, Jurgen. 1987a. *The Philosophical Discourse of Modernity: Twelve Lectures*. Cambridge, MA: MIT Press.

Habermas, Jurgen. 1987b. *The Theory of Communicative Action: Vol. 2*. Boston: Beacon Press.

Habermas, Jurgen. 1989. *The Structural Transformation of the Public Sphere*. Cambridge, MA: MIT Press.

Hale, C.B. 1990. *Infant Mortality: An American Tragedy*. Washington, DC: Population Reference Bureau Inc.

Hall, G. Stanley. 1904. *Adolescence: Its Psychology and its Relations to Physiology, Anthropology, Sociology, Sex, Crime, Religion and Education*. New York: D. Appleton and Company.

Hall, G. Stanley. 1909. *Fifty Years of Darwinism: Modern Aspects of Evolution*. Centennial addresses in honor of Charles Darwin, before the American Association for the Advancement of Science, Baltimore, Friday, January 1, 1909. New York: H. Holt and Company.

Haraway, Donna. 1991. *Simians, Cyborgs and Women: The Reinvention of Nature*. London: Routledge.

Harley, J. Brian. 1992. Deconstructing the map. In Trevor J. Barnes and James S. Duncan (eds) *Writing Worlds: Discourse, Text and Metaphor in the Representation of Landscape*, pp. 231–47. London: Routledge.

Hart, Roger A. 1979. *Children's Experience of Place*. New York: Irvington.

Hart, Roger A. 1984. The geography of children and children's geography. In T.F. Saarinen, D. Seamon and J.L. Sell (eds) *Environmental Perception and Behavior: An Inventory and Prospect*, pp. 99–129. Chicago: University of Chicago Press.

Hart, Roger A. 1997. *Children's Participation: The Theory and Practice of Involving Young Citizens in Community Development and Environmental Care*. London: UNICEF/Earthscan Publications Ltd.

Hart, Roger A. and Gary T. Moore. 1973. The development of spatial cognition: a review. Roger M. Downs and David Stea (eds) *Image and Environment*, pp. 246–88. Chicago: Aldine.

Harvey, David. 1992. Postmodern morality plays. *Antipode* 24: 300–26.

Harvey, David. 1993. From space to place and back again: reflections on the condition of postmodernity. In Jon Bird, Barry Curtis, Tim Putnam, George Robertson and Lisa Tickner (eds) *Mapping the Futures: Local Cultures, Global Change*, pp. 3–29. London: Routledge.

Hendrick, Harry. 1990. Constructions and reconstructions of childhood: an interpretative survey from 1800 to the present. In A. James and A. Prout (eds) *Constructing and Reconstructing Childhood: Contemporary Issues in the Sociological Study of Childhood*. London: Falmer Press.

Hengst, Herbert. 1987. The liquidation of childhood – an objective tendency. *International Journal of Sociology* 17: 58–80.

Henriques, Julian, Wendy Hollway, Cathy Urwin, Couze Venn and Valerie Walkerdine. 1984. *Changing the Subject: Psychology, Social Regulation and Subjectivity*. London: Methuen.

Herod, Andrew. 1991. The production of scale in the United States labour relations. *Area* 23(1): 82–8.

Holloway, Sarah L. 1998a. Local childcare cultures: moral geographies of mothering and the social organization of pre-school education. *Gender, Place and Culture* 5(1): 29–53.

Holloway, Sarah L. 1998b. Geographies of justice: preschool-childcare provision and the conceptualisation of social justice. *Environment and Planning C* 16: 85–104.

Holloway, Sarah L. and Gill Valentine (eds). 2000a. *Children's Geographies: Playing, Living, Learning.* London and New York: Routledge.

Holloway, Sarah L. and Gill Valentine (eds). 2000b. Children's geographies and the new social studies of childhood. In Sarah L. Holloway and Gill Valentine (eds) *Children's Geographies: Playing, Living, Learning*, pp. 1–28. London and New York: Routledge.

Holloway, Sarah L., Gill Valentine and Nick Bingham. 2000. Institutionalising technologies: masculinities, femininities, and the heterosexual economy of the IT classroom. *Environment and Planning A* 32(4): 617–34.

Honneth, Alex. 1995. *The Struggle for Recognition: The Moral Grammar of Social Conflicts.* Trans. Joel Anderson. Cambridge: Polity Press.

Hyams, Melissa S. 2000. "Pay attention in class . . . [and] don't get pregnant": a discourse of academic success among adolescent Latinas. *Environment and Planning A* 32(4): 617–35.

Irigaray, Luce. 1985. *The Sex Which Is Not One.* Trans. Gillian C. Gill. Ithaca, NY: Cornell University Press.

Irigaray, Luce. 1993. *An Ethics of Sexual Difference.* Trans. Carolyn Burke and Gillian C. Gill. Ithaca, NY: Cornell University Press.

Ivy, Marilyn. 1995. Have you seen me? Recovering the inner child in late twentieth-century America. In Sharon Stephens (ed.) *Children and the Politics of Culture*, pp. 79–104. Princeton, NJ: Princeton University Press.

Jablonsky, Thomas. 1993. *Pride in the Jungle: Community and Everyday Life in Back of the Yards Chicago.* Baltimore, MD: Johns Hopkins University Press.

James, Allison. 1993. *Childhood Identities: Self and Social Relationships in the Experience of the Child.* Edinburgh: Edinburgh University Press.

James, Alison, Chris Jenks and Alan Prout. 1998. *Theorizing Childhood.* New York: Teachers' College Press.

James, Susan. 1990. Is there a "place" for children in geography. *Area* 22(3): 278–83.

Jameson, Fredric. 1984. Postmodernism, or the cultural logic of late capitalism. *New Left Review* 146: 53–92.

Jameson, Fredric. 1992. *The Geopolitical Aesthetic: Cinema and Space in the World System.* Bloomington: Indiana University Press.

Jenks, Chris. 1982. Constructing the child. In C. Jenks (ed.) *The Sociology of Childhood: Essential Readings.* London: Batsford.

Jenks, Chris. 1996. *Childhood.* London: Routledge.

Jonas, Andrew. 1994. Editorial: the scale politics of spatiality. *Environment and Planning D: Society and Space* 12(3): 257–64.

Jordanova, Ludmilla. 1989. Children in history: concepts of nature and society. In Geoffrey Scarre (ed.) *Children, Parents and Politics*, pp. 3–24. Cambridge: Cambridge University Press.

Katz, Cindi. 1986. Children and the environment: work, play and learning in rural Sudan. *Children's Environments Quarterly* 3(4): 43–51.

Katz, Cindi. 1991a. A cable to cross a curse: the everyday practices of resistance and reproduction among youth in New York City. City University of New York: Department of Environmental Psychology.

Katz, Cindi. 1991b. Sow what you know: the struggle for social reproduction in rural Sudan. *Annals of the Association of American Geographers* 81(3): 488–514.

Katz, Cindi. 1993. Growing girls/closing circles. In Cindi Katz and Jan Monk (eds) *Full Circles: Geographies of Women over the Lifecourse*, pp. 88–106. London: Routledge.

Katz, Cindi. 1994. Playing the field: questions of fieldwork in geography. *The Professional Geographer* 46(1): 67–72.

Katz, Cindi. 1998. Disintegrating developments: global economic restructuring and the eroding of ecologies of youth. In Tracey Skelton and Gill Valentine (eds) *Cool Places: Geographies of Youth Cultures*, pp. 130–44. London and New York: Routledge.

Katz, Cindi. 2001. Stuck in place: the local consequences of the globalization of social reproduction. *Antipode*. In R.J. Johnston, Peter R. Taylor and Michael Watts (eds) *Geographies of Global Change*, 2nd edn. Oxford: Blackwell. Forthcoming.

Katz, Cindi and Andrew Kirby. 1991. In the nature of things: the environment and everyday life. *Transactions, Institute of British Geographers*, New Series 16(3): 259–71.

Kincaid, James R. 1992. *Child-Loving: The Erotic Child and Victorian Literature.* New York: Routledge.

King, Michael. 1999. *Moral Agendas for Children's Welfare.* London and New York: Routledge.

Kirby, Kathleen. 1996. *Indifferent Boundaries: Spatial Concepts of Human Subjectivity*. New York: Guilford Press.

Kobayashi, Audrey and Linda Peake. 2000. Racism out of place: thoughts on whiteness and an antiracist geography in the new millennium. *Annals of the Association of American Geographers* 90(2): 392–403.

Kristeva, Julia. 1982. *Power of Horrors*. New York: Columbia University Press.

Kulynych, Jessica, J. 1999. No playing in the public sphere: democratic theory and the exclusion of children. Paper presented at the Annual Meeting of the American Political Science Association, Atlanta.

Lacan, Jacques. 1978. *The Four Fundamental Concepts of Psychoanalysis*. New York: Norton.

Laqueur, Thomas W. 1990. *Making Sex: Body and Gender from the Greeks to Freud*. Cambridge, MA: Harvard University Press.

Lefebvre, Henri. 1991. *The Production of Space*. Oxford: Blackwell.

Liben, Lynn S. 1981. Spatial representation and behavior: multiple perspectives. In Lynn S. Liben, A.H. Patterson and N. Newcombe (eds) *Spatial Representation and Behavior Across the Lifespan*. New York: Academic Press.

Liben, Lynn S. and Roger M. Downs. 1989. Understanding maps as symbols: the development of map concepts in children. In H.W. Reese (ed.) *Advances in Child Development and Behavior*, Vol. 22, pp. 145–201. New York: Academic Press.

Liben, Lynn S. and Roger M. Downs. 1997. Can-ism and can'tianism: a straw child. *Annals of the Association of American Geographers*: 159–67.

Livingstone, David N. 1992. *The Geographical Tradition*. Oxford: Blackwell.

Longhurst, Robyn. 1995. The body and geography. *Gender, Place and Culture* 2: 97–106.

Longhurst, Robyn. 1997. (Dis)embodied geographies. *Progress in Human Geography* 21(4): 486–501.

Lucas, Tim. 1998. Youth gangs and moral panics in Santa Cruz, California. In Tracey Skelton and Gill Valentine (eds) *Cool Places: Geographies of Youth Cultures*, pp. 145–60. London and New York: Routledge.

Lulka, David. 2000. Review of J. Wolch and J. Emel (eds) *Animal Geographies: Place, Politics and Identity in the Nature–Culture Borderlands*. *The Professional Geographer*, 52(3): 589–90.

Lynch, Kevin. 1960. *The Image of the City*. Cambridge, MA: MIT Press.

Lynch, Kevin. 1977. *Growing Up in Cities*. Cambridge, MA: MIT Press.

McClintock, Anne. 1995. *Imperial Leather*. London and New York: Routledge.

McDowell, Linda. 2000. The trouble with men? Young people, gender transformations and the crisis of masculinity. *International Journal of Urban and Regional Research* 24: 201–9.

McDowell, Linda. 2001. Learning to serve? Employment aspirations and attitudes of young men in an era of labor market restructuring. *Gender, Place and Culture* 7(4).

McKechnie, George E. 1974. *Manual for the Environmental Response Inventory*. Palo Alto, CA: Consulting Psychological Press.

McKechnie, George E. 1977. The environmental response inventory in application. *Environment and Behavior* 9: 255–76.

McKendrick, John H., Michael G. Bradford and Anna V. Fielder. 2000. Time for a party!: making sense of the commercialization of leisure space for children. In Sarah Holloway and Gill Valentine (eds) *Children's Geographies: Playing, Living and Learning*, pp. 100–18. London and New York: Routledge.

McKenzie, B.E., R.H. Day and E. Ihsen. 1984. Localization of events in space: young children are not always egocentric. *British Journal of Developmental Psychology* 2: 1–9.

McLaren, Peter. 1989. *Life In Schools: An Introduction to Critical Pedagogy in the Foundations of Education*. New York: Longman.

McLaren, Peter. 1995. *Critical Pedagogy and Predatory Culture*. London and New York: Routledge.

McLaren, Peter and Rhonda Hammer. 1996. Media knowledges, warrior citizenry, and postmodern literacies. In Henry Giroux, Colin Lankshear, Peter McLaren and Michael Peters (eds) *Counternarratives: Critical Studies and Critical Pedagogies in Postmodern Spaces*, pp. 81–117. London and New York: Routledge.Males, Mike A. 1997. Debunking 10 myths about teens. *The Education Digest* 63(4): 48–53.

Males, Mike A. 2000. Punishing teens to protect them. *Los Angeles Times*, 11 June, M–1 and M–3.

Marcus, George E. 1986. Contemporary problems of ethnography in the modern world system. In James Clifford and George E. Marcus (eds) *Writing Culture: The Poetics and Politics of Ethnography*, pp. 165–93. Berkeley: University of California Press.

Marx, Karl. 1976. *Capital*, Vol. 1. Trans. B. Fowkes. Harmondsworth: Penguin Books.

Massey, Doreen. 1993. Power-geometry and a progressive sense of place. In Jon Bird, Barry Curtis, Tim Putnam, George Robertson and Lisa Tickner (eds) *Mapping the Futures: Local Cultures, Global Change*, pp. 59–69. London and New York: Routledge.

Massey, Doreen. 1994. *Space, Place and Gender*. Minnesota: University of Minnesota Press.

Matthews, Hugh M. 1984a. Cognitive maps: a comparison of graphic and iconic techniques. *Area* 16: 33–40.

Matthews, Hugh M. 1984b. Environmental cognition of young children: images of school and home area. *Transactions of the Institute of British Geographers* 9(1): 89–105.

Matthews, Hugh M. 1986. Children as map makers. *The Geographical Magazine*, August: 47–9.

Matthews, Hugh M. 1992. *Making Sense of Place*. Lanham, MD: Barnes and Noble.

Matthews, Hugh and Melanie Limb. 1999. Defining an agenda for the geography of children: review and prospect. *Progress in Human Geography* 23(1): 61–90.

Matthews, Hugh, Melanie Limb and Mark Taylor. 2000. The "street as thirdspace." In Sarah L. Holloway and Gill Valentine (eds) *Children's Geographies: Playing, Living and Learning*, pp. 119–38. London and New York: Routledge.

Menaker, Esther. 1995. *The Freedom to Inquire*. Northvale, NJ: Jaron Aronson.

Merrifield, Andy. 1995. Situated knowledge through exploration: reflections on Bunge's "Geographical expeditions." *Antipode* 27(1): 49–70.

Mitchell, Don. 1993. Public housing in single-industry towns: changing landscapes of paternalism. In James Duncan and David Ley (eds) *Place/Culture/Representation*, pp. 110–27. London and New York: Routledge.

Montagu, Ashley. 1971. *The Elephant Man: A Study in Human Dignity*. New York: Outerbridge & Dienstfrey, distributed by E. P. Dutton.

Moore, Gary T. 1976. Theory and research on the development of environmental knowing. In Gary T. Moore and Reginald G. Golledge (eds), pp. 83–107. Strowdsburg, PA: Dowden, Hutchinson and Ross.

Moore, Robin. 1986. *Childhood's Domain: Place and Play in Child Development*. London: Croom Helm.

Morris, Debra. 2000. Privacy, privation, perversity: toward new representations of the personal. *Signs: Journal of Women in Culture and Society* 25(2): 324–51.

Morrison, Toni (with Slade Morrison). 1999. *The Big Box*. New York: Hyperion Books.

Morrow, V. 1994. Responsible children? Aspects of children's work and employment outside of school in contemporary UK. In B. Mayall (ed.) *Children's Childhoods: Observed and Experienced*. London: Falmer.

Morss, J. 1990. *The Biologising of Childhood*. Hillsdale, NJ: Erlbaum.

Nabhan, Gary Paul and Stephen Trimble. 1994. *The Geography of Childhood: Why Children Need Wild Places*. Boston: Beacon Press.

Nagel, Thomas. 1986. *A View from Nowhere*. New York: Firebrand Books.

Nast, Heidi. 1994. Women in the field: critical feminist methodologies and theoretical perspectives. *The Professional Geographer* 46(1): 54–66.

Nast, Heidi. 1998. The body as "place." In Heidi Nast and Steve Pile (eds) *Places Through the Body*, pp. 93–116. New York and London: Routledge.

Nast, Heidi. 2000. Mapping the "unconscious": racism and the Oedipal family. *Annals of the Association of American Geographers* 90(2): 215–55.

Nast, Heidi and Steve Pile. 1998. *Places Through the Body*. New York and London: Routledge.

Natter, Wolfgang and J.P. Jones. 1993. Pets or meat: class, ideology and space. *Roger & Me. Antipode* 25(2): 140–58.

Nelson, Lise. 1999. Bodies (and spaces) do matter: the limits of performativity. *Gender, Place and Culture* 6(4): 331–53.

Ng, S.H. 1983. Children's ideas about banking and shop profit. *Journal of Econmic Psychology* 4: 209–21.

Nieuwenhuys, Olga. 1994. *Children's Lifeworlds: Gender, Welfare and Labor in the Developing World*. London and New York: Routledge.

Ogbu, J.U. 1987. Variability in minority school performance: a problem in search of an explanation. *Anthropology and Educational Quarterly* 18(4): 312–32.

O'Keefe, J. and L. Nadel. 1978. *The Hippocampus as a Cognitive Map*. Oxford: Clarendon Press.

Okin, Susan Moller. 1989. *Justice, Gender and the Family*. New York: Basic Books.

Oldman, David. 1994. Adult–child relations as class relations. In Jens Qvortrup, Marjatta Bardy, Giovanni Sgritta and Helmut Wintersberger (eds) *Childhood Matters: Social Theory, Practice and Politics*, pp. 43–58. Aldershot, UK: Avebury Press.

Phillips, M.H., D. Kronenfeld and V. Jeter. 1986. A model of services to homeless families in shelters. In J. Erickson and C. Wilhelm (eds) *Housing the Homeless*, pp. 322–34. New Brunswick, NJ: Center for Urban Policy Research.

Piaget, Jean. 1952. *The Origins of Intelligence in Children*. New York: Harcourt, Brace and World, Inc.

Piaget, Jean. 1954. *The Construction of Reality in the Child*. New York: Harcourt, Brace and World, Inc.

Piaget, Jean. 1971. *Structuralism*. New York: Basic Books.

Piaget, Jean and Brenda Inhelder. 1956. *The Child's Conception of Space*. London: Routledge & Kegan Paul.

Pick, H.L. 1976. Transactional-constructivist approach to environmental knowing. In G.T. Moore and R.G. Golledge (eds) *Environmental Knowing*, pp. 185–8. Stroudsburg, PA: Dowden, Hutchinson and Ross.

Pile, Steve. 1993. Human agency and human geography revisited: a critique of "new models" of the self. *Transactions of the Institute of British Geographers* 18(1): 122–39.

Pile, Steve. 1996. *The Body and the City: Psychoanalysis, Space and Subjectivity*. London and New York: Routledge.

Pile, Steve and Nigel Thrift. 1995. *Mapping the Subject: Geographies of Cultural Transformation*. New York and London: Routledge.

Plumwood, Vera. 1992. Feminism and ecofeminism: beyond dualistic assumptions of women. *Ecologist* 22(1): 8–13.

Pollock, Linda. 1983. *Forgotten Children: Parent–Child Relations, 1500–1900*. Cambridge: Cambridge University Press.

202

Popenoe, David. 1993. American family decline, 1960–1990: a review and appraisal. *Journal of Marriage and Family* 55: 527–55.

Postman, Neil. 1982. *The Disappearance of Childhood*. New York: Delacourt Press.

Putnam, Robert D. 1993. The prosperous community: social capital and public life. *The American Prospect* 13 (Spring): 37.

Quetelet, Lambert J. 1835. *Physique social. Essai sur le developpement des facultés de l'homme, 2* vols. Brussels: Muquardt.

Qvortrup, Jens. 1985. Placing children in the division of labour. In Paul Close and Rosemary Collins (eds) *Family and Economy in Modern Society*, pp. 129–45. London: Macmillan.

Qvortrup, Jens, M. Bardy, G. Sgritta and H. Winterberger (eds). 1994. *Childhood Matters*. Aldershot, UK: Avebury Press.

Rawls, John. 1971. *A Theory of Justice*. Oxford: Oxford University Press.

Richardson, D. 1993. *Women, Motherhood and Childrearing*. London: Macmillan.

Riis, Jacob. 1890/1997. *How the Other Half Lives: Studies Among the Tenements of New York*. New York: Penguin Books.

Rivlin, Leane G. and Maximem Wolfe. 1985. *Institutional Settings in Children's Lives*. New York: Wiley.

Roberts, Sue. 1998. Commentary: what about children? *Environment and Planning A* 30: 3–11.

Robertson, P. 1976. Home as a nest: middle class childhood in nineteenth century Europe. In L. De Mause (ed.) *The History of Childhood*. London: Souvenir.

Robins, Kevin. 1996. *Into the Image: Culture and Politics in the Field of Vision*. London and New York: Routledge.

Robson, Elsbeth. 1996. Working girls and boys: children's contributions to household survival in West Africa. *Geography* 81: 403–7.

Robson, Elsbeth. 2001. Hidden child workers: carers in Zimbabwe. *Antipode*. In review.

Robson, Elsbeth and Nicola Ansell. 2000. Young carers in South Africa: exploring stories from Zimbabwean secondary school students. In Sarah L. Holloway and Gill Valentine (eds) *Children's Geographies: Playing, Living and Learning*, pp. 174–93. London and New York: Routledge.

Rodgers, G. and G. Standing. 1981. *Child Work, Poverty and Underdevelopment*. Los Angeles and Berkeley: University of California Press.

Rose, Gillian. 1993. *Feminism and Geography*. Minnesota: University of Minnesota Press.

Rosser, Susan. 1991. Eco-feminism: lessons from feminism and ecology. *Women's Studies International Forum* 14(3): 143–51.

Rousseau, Jean-Jacques. (1962) *The Emile of Jean-Jacques Rousseau: Selections*. Translated and edited by William Boyd. New York: Columbia University, Bureau of Publications.

Ruddick, Susan. 1995. *Young and Homeless in Hollywood: Mapping Social Identities*. New York and London: Routledge.

Ruddick, Susan. 1998. Modernism and resistance: how ''homeless'' youth sub-cultures make a difference. In Tracey Skelton and Gill Valentine (eds) *Cool Places: Geographies of Youth Cultures*, pp. 343–60. London and New York: Routledge.

Ruffy, M. 1981. Influence of social factors in the development of the young child's moral judgement. *European Journal of Social Psychology* 11: 61–75.

Sack, Robert D. 1999. A sketch of a geographic theory of morality. *Annals of the Association of American Geographers* 89(1): 26–44.

Samuels, Marwyn. 1978. Existentialism and human geography. In D. Ley and M. Samuels (eds) *Humanistic Geography: Prospects and Problems*, pp. 22–40. London: Croom Helm.

Sanders, Rickie and Mark T. Mattson. 1998. *Growing Up in America: An Atlas of Youth in the USA.* New York: Simon & Schuster Macmillan.

Saunders, Ralph. 1999. The pace community policing makes and the body that makes it. *The Professional Geographer* 51(1): 135–46.

Shaw, David. 2000. Kids are people too, papers decide. *Los Angeles Times*, July 11, A1, col. 1.

Sibley, David. 1991. Children's geographies: some problems of representation. *Area* 23(3): 269–70.

Sibley, David. 1995a. *Geographies of Exclusion.* London and New York: Routledge.

Sibley, David. 1995b. Families and domestic routines: constructing the boundaries of childhood. In Steve Pile and Nigel Thrift (eds) *Mapping the Subject: Geographies of Cultural Transformation*, pp. 123–37. London: Routledge.

Siegel, Andrew W. 1981. The externalization of cognitive maps by children and adults: in search of ways to ask better questions. In L.S. Liben, A.H. Patterson and N. New-combe (eds) *Spatial Representation and Behavior Across the Lifespan.* New York: Academic Press.

Simonsen, Kirsten. 2000. Editorial: the body as battlefield. *Transactions of the Institute of British Geographers* 25: 7–9.

Skelton, Tracey. 2000. Nothing to do, nowhere to go?: Teenage girls and ''public space'' in the Rhondda Valleys, South Wales. In Sarah L. Holloway and Gill Valentine (eds) *Children's Geographies: Playing, Living and Learning*, pp. 80–99. London and New York: Routledge.

Skelton, Tracey and Gill Valentine (eds). 1997. *Geographies of Youth Cultures.* London and New York: Routledge.

Smiles, Samuel. 1838. *Physical Education; or, the Nurture and Management of Children, Founded on the Study of their Nature and Constitution.* Edinburgh: Oliver & Boyd.

Smith, Neil. 1996. *The New Urban Frontier: Gentrification and the Revanchist City.* London and New York: Routledge.

Smolensky, E., S. Danziger and P. Gottschalk. 1988. The declining significance of age in the United States: trends in the well-being of children and the elderly since 1939. In J.L. Palmer, T. Smeeding and B.B. Torrey (eds) *The Vulnerable.* Washington, DC: The Urban Institute Press.

Spencer, Chris and Mark Blades. 1986. Pattern and process: a review essay on the relationship between behavioral geography and environmental psychology. *Progress in Human Geography* 10(2): 230–48.

Spencer, Chris, Mark Blades and K. Morsley. 1989. *The Child in the Physical Environment: The Development of Spatial Knowledge and Cognition.* Chichester: John Wiley.

Stainton-Rogers, Rex. 1989. The social construction of childhood. In W. Stainton-Rogers, D. Harvey and E. Ash (eds) *Child Abuse and Neglect.* London: Open University Press.

Stainton-Rogers, Wendy and Rex Stainton-Rogers. 1992. *Stories of Childhood: Shifting Agendas of Child Concern.* London: Open University Press.

Stainton-Rogers, Wendy and Rex Stainton-Rogers. 1999. What is good and bad sex for children? In Michael King (ed.) *Moral Agendas for Children's Welfare*, pp. 179–97. London and New York: Routledge.

Steedman, Carolyn. 1995. *Strange Dislocations: Childhood and the Idea of Human Interiority, 1780–1930.* Cambridge, MA: Harvard University Press.

Stephens, Sharon. 1995. Children and the politics of culture in "late capitalism." In Sharon Stephens (ed.) *Children and the Politics of Culture*, pp. 3–48. Princeton, NJ: Princeton University Press.

Stone, Lawrence. 1974. The massacre of the innocents. *New York Review of Books* 21(18): 27, November.

Temple, William. 1992. *The American Heritage Dictionary of the English Language*, 3rd edn. Boston: Houghton Mifflin Company.

Thoreau, Henry David. 1993, first published in 1854. *Walden and Other Writings.* New York: Barnes & Noble.

Thorne, Barrie. 1994. *Gender Play: Girls and Boys in School.* New Brunswick, NJ: Rutgers University Press.

Tolman, Edward. 1948. Cognitive maps in rats and men. *Psychological Review* 55: 189–208.

Trumbach, Randolph. 1999. London. In David Higgs (ed.) *Queer Sites: Gay Urban Histories Since 1600*, pp. 89–111. London and New York: Routledge.

Tversky, Barbara, Julie Bauer Morrison, Nancy Franklin and David J. Bryant. 1999. Three spaces of spatial cognition. *The Professional Geographer* 51(4): 516–24.

United States Department of Health and Human Services. 1986. *Infant Mortality and Low Birth Weight.* Washington, DC: United States Government Printing Office.

Urwin, Cathy. 1984. Power relations and the emergence of language. In Julian Enriques, Wendy Hollway, Cathy Urwin, Couze Venn and Valerie Walkerdine (eds) *Changing the Subject: Psychology, Social Regulation and Subjectivity*, pp. 264–322. London and New York: Methuen.

Valentine, Gill. 1995. Creating transgressive space: the music of KD Lang. *Transactions of the Institute of British Geographers* 20(4): 474–87.

Valentine, Gill. 1996. Angels and devils: moral landscapes of childhood. *Environment and Planning D: Society and Space* 14: 581–99.

Valentine, Gill. 1997a. "Oh yes I can," "Oh no you can't": children and parents' understandings of kids' competence to negotiate public space safely. *Antipode* 29(1): 65–89.

Valentine, Gill. 1997b. A safe place to grow up? Parenting, perceptions of children's safety and the rural idyll. *Journal of Rural Studies* 13(2): 137–48.

Valentine, Gill. 1997c. "My son's a bit dizzy" "My wife's a bit soft": gender, children and cultures of parenting. *Gender, Place and Culture* 4: 37–62.

Valentine, Gill and Sarah L. Holloway. 2001. On-line dangers?: Geographies of parents' fears for children's safety in cyberspace. *The Professional Geographer* 53(1): 71–83.

Valentine, Gill, Tracey Skelton and Deborah Chambers. 1998. Cool places: an introduction to youth and youth cultures. In Tracey Skelton and Gill Valentine (eds) *Cool Places: Geographies of Youth Cultures*, pp. 1–32. London and New York: Routledge.

Valentine, Gill, Sarah L. Holloway and Nick Bingham. 2000. Transforming cyberspace: children's interventions in the new public sphere. In Sarah L. Holloway and Gill Valentine (eds) *Children's Geographies: Playing, Living and Learning*, pp. 156–73. London and New York: Routledge.

Vygotsky, Lev S. 1987. Thinking and its development in childhood. In R.W. Rieber and A.S. Carton (eds) *The Collected Works of L.S. Vygotsky, Vol 1: Problems of General Psychology*. New York: Plenum Press.

Walkerdine, Valerie. 1984. Developmental psychology and the child-centred pedagogy: the insertion of Piaget into early education. In Julian Enriques, Wendy Hollway, Cathy Urwin, Couze Venn and Valerie Walkerdine (eds) *Changing the Subject: Psychology, Social Regulation and Subjectivity*, pp.153–202. London and New York: Methuen.

Walkerdine, Valerie. 1988. *The Mastery of Reason: Cognitive Development and the Production of Rationality*. London and New York: Routledge.

Walmsley, James D. 1988. *Urban Living: The Individual in the City*. London: Longman.

Ward, Colin. 1978. *The Child in the City*. London: Architectural Press.

Ward, Colin. 1988. *The Child in the Country*. London: Bedford Square Press.

Warren, K. 1987. Feminism and ecology: making connections. *Environmental Ethics* 9: 3–20.

Warren, K. 1990. The power and promise of ecological feminism. *Environmental Ethics* 12: 125–46.

Wertsch, James. 1985. Introduction. In James Wertsch (ed.) *Cultural Communication and Cognition: Vygotskian Perspectives*, pp. 1–17. Cambridge: Cambridge University Press.

White, Morton and Lucia White. 1962. *The Intellectual Versus the City: From Thomas Jefferson to Frank Lloyd Wright*. Cambridge, MA: Harvard University Press.

Willis, Paul. 1981, first published 1977. *Learning to Labour: How Working Class Kids Get Working Class Jobs*. New York: Columbia University Press.

Wilson, Adrian. 1980. The infancy of the history of childhood: an appraisal of Philippe Ariès. *History and Theory* 19(2): 132–54.

Winchester, Hilary. 1991. The geography of children. *Area* 23(4): 357–60.

Winnicott, Donald W. 1965. *The Family and Individual Development*. New York: Basic Books.

Winnicott, Donald W. 1971. *Playing and Reality*. London: Tavistock.

Winnicott, Donald W. 1975. *Through Paediatrics to Psycho-analysis*. New York: Basic Books.

Winnicott, Donald W. 1988. *The Child, The Family and the Outside World*. London: Penguin

Wolch, Jennifer and Jody Emel. 1998. *Animal Geographies: Place, Politics and Identity in the Nature–Culture Borderlands*. London and New York: Verso.

Wolfe, Maxime and Leanne G. Rivlin. 1987. The institutions in children's lives. In C.S. Weinstein and T.G. David (eds) *Spaces for Children: The Built Environment and Child Development*, pp. 89–116. New York: Plenum.

Wood, Denis. 1982. To catch the wind. *Outlook* 46: 3–31.

Wood, Denis. 1985a. Doing nothing. *Outlook* 57: 3–20.

Wood, Denis. 1985b. Nothing doing. *Children's Environments Quarterly* 2(2): 14–25.

Wood, Denis. 1993. *The Power of Maps*. London and New York: Routledge.

Wood, Denis and Robert Beck. 1990. Dos and don'ts: family rules, rooms and their relationships. *Children's Environments Quarterly* 7(1): 2–14.

Wood, Denis and Robert Beck. 1994. *The Home Rules*. Baltimore, MD, and London: Johns Hopkins University Press.

Yelland, Nicola and Susan Grieshaber. 1998. Blurring the edges. In Nicola Yelland (ed.) *Gender in Early Childhood*, pp. 1–14. New York and London: Routledge.

Young, Iris Marion. 1990a. The ideal of community and the politics of difference. In Linda Nicholson (ed.) *Feminism/Postmodernism*, pp. 300–23. London and New York: Routledge.

Young, Iris Marion. 1990b. *Justice and the Politics of Difference*. Princeton, NJ: Princeton University Press.

Yuval-Davis, Nira. 1993. Gender and nation. *Racial and Ethnic Studies* 16(4): 621–32.

Zipes, Jack. 1997. *Happily Ever After: Fairy Tales, Children, and the Culture Industry*. New York and London: Routledge.

Zita, Jacquelyn N. 1998. *Body Talk: Philosophical Reflections on Sex and Gender*. New York: Columbia University Press.

Zonn, Leo E. and Stuart C. Aitken. 1994. Of pelicans and men: symbolic landscapes, gender and Australia's *Storm Boy*. In S.C. Aitken and L.E. Zonn (eds) *Place, Power, Situation and Spectacle: A Geography of Film*, pp. 137–59. Lanham, MD: Rowman and Littlefield.

INDEX

adolescence: and class 56–7; construction of 5–6, 77, 129; and disembeddedness 155–6; pregnancy 77–9, 135–40; as wild 34–6
Ainsworth, Mary 95, 96
alterity *see* difference
Anzaldúa, Gloria 88, 109–10, 115
Appleton, Jay 42–3, 60n
Archard, David 40, 93, 175
Ariès, Phillipe 80, 120–5, 130, 166
Armstrong, David 71–2
authority 54, 132, 173–6

Beazley, Harriott 6, 160–1
biologism 28–30, 69–71, 95, 96
Blaut, James 14, 27, 47–9
Blum, Virginia 84–5, 98–102
body, constructions of: as activity 109–10, 115; adolescent 77–9; adult/child distinctions 70–1, 89; biological 69, 109; cancer 76; clothing 77–9, 81, 139; defined 65; and developmental theory 66, 75–6; in geography 62, 65–8; horticultural metaphor 70–1; ideal forms 69–70, 75–6; in Lefebvre 100–2; *mignontage* 79–84, 123–4; mind/body split 67–8; norms 67, 74–6; as other 66–7; phallocentrism 98–102; scientific 29–30, 68–71; technology 83–4; *see also* childhood, constructions of; sexuality
body, control of: circumcision 73–4, 87n, 117n; coddling 70, 80, 123; colonial 105–9; playground surveillance 125–8; the racist oedipal family 107, 136–7; rules 17–18; teenage pregnancy 77–9, 135–40; toilet training 66–7; *see also* sexuality; space

Bowlby, John 95, 96
Bradshaw, John 58–9
Bulger, Jamie, murder of 146–7, 167
Bunge, William 12–14, 23
Bunting, Trudi 43
Burman, Erica 44
Butler, Judith 102–4

Callard, Felicity 83–4
capital, social and cultural 135–40, 150, 154–5, 159
capitalism 83–4, 142–3; *see also* commodification; globalization
Capital (Marx) 84
caregivers 66–7, 75
Cartesianism: cognitive maps 50–1; and identity theory 100–1; mind/body split 67–8, 69, 101, 113; and resistance 80, 109–10; subjectivity 22, 49–50
cartography *see* mapping
Centuries of Childhood (Ariès) 121–5, 166
child care 137–8, 150–4
childhood, constructions of: commodified 129–30, 148–50; globalized 10–12, 142–4; historical 120–5, 166; individualization 129–30; *mignontage* 79–84, 123–4; modern 64, 130, 143, 166; otherization 7–8, 126; peer group 131–5; post-structural 57–60; as problematic 5–7, 21, 122; unchildlike behavior 146–9, 158–61, 162–8; wildness 31–6, 32–8, 105, 127–8, 158; *see also* adolescence; developmental theory; identity; moralities
circumcision 73–4, 83, 87n, 117n
Cisneros, Sandra 49
class 56–7, 132, 135–6
classrooms 53–5

208